Internationalization OF THE Nuclear Fuel Cycle

GOALS, STRATEGIES, AND CHALLENGES

U. S. Committee on the
Internationalization of the Civilian Nuclear Fuel Cycle

Committee on International Security and Arms Control,
Policy and Global Affairs

Nuclear and Radiation Studies Board
Division on Earth and Life Sciences

NATIONAL ACADEMY OF SCIENCES *AND*
NATIONAL RESEARCH COUNCIL
OF THE NATIONAL ACADEMIES

Russian Committee on the
Internationalization of the Civilian Nuclear Fuel Cycle

RUSSIAN ACADEMY OF SCIENCES

THE NATIONAL ACADEMIES PRESS
Washington, D.C.
www.nap.edu

THE NATIONAL ACADEMIES PRESS 500 Fifth Street, N.W. Washington, DC 20001

NOTICE: The project that is the subject of this report was approved by the Governing Board of the National Research Council, whose members are drawn from the councils of the National Academy of Sciences, the National Academy of Engineering, and the Institute of Medicine. The members of the committee responsible for the report were chosen for their special competences and with regard for appropriate balance.

This study was supported by Grants from the John D. and Catherine T. MacArthur Foundation and the Carnegie Corporation of New York. Additional support was provided by the Russian Academy of Sciences. Any opinions, findings, conclusions, or recommendations expressed in this publication are those of the author(s) and do not necessarily reflect the views of the organizations or agencies that provided support for the project.

International Standard Book Number: 13-978-0-309-12660-1
Library of Congress Catalog Card Number: 10-0-309-12660-6

A limited number of complimentary copies are available from the Committee on International Security and Arms Control, National Academy of Sciences, 500 Fifth Street, N.W., Washington, DC 20001; +202-334-2811.

Additional copies of this report are available from the National Academies Press, 500 Fifth Street, N.W., Lockbox 285, Washington, DC 20055; (800) 624-6242 or (202) 334-3313 (in the Washington metropolitan area); Internet, *http://www.nap.edu.*

Printed in the United States of America.

THE NATIONAL ACADEMIES

Advisers to the Nation on Science, Engineering, and Medicine

The **National Academy of Sciences** is a private, nonprofit, self-perpetuating society of distinguished scholars engaged in scientific and engineering research, dedicated to the furtherance of science and technology and to their use for the general welfare. Upon the authority of the charter granted to it by the Congress in 1863, the Academy has a mandate that requires it to advise the federal government on scientific and technical matters. Dr. Ralph J. Cicerone is president of the National Academy of Sciences.

The **National Academy of Engineering** was established in 1964, under the charter of the National Academy of Sciences, as a parallel organization of outstanding engineers. It is autonomous in its administration and in the selection of its members, sharing with the National Academy of Sciences the responsibility for advising the federal government. The National Academy of Engineering also sponsors engineering programs aimed at meeting national needs, encourages education and research, and recognizes the superior achievements of engineers. Dr. Charles M. Vest is president of the National Academy of Engineering.

The **Institute of Medicine** was established in 1970 by the National Academy of Sciences to secure the services of eminent members of appropriate professions in the examination of policy matters pertaining to the health of the public. The Institute acts under the responsibility given to the National Academy of Sciences by its congressional charter to be an adviser to the federal government and, upon its own initiative, to identify issues of medical care, research, and education. Dr. Harvey V. Fineberg is president of the Institute of Medicine.

The **National Research Council** was organized by the National Academy of Sciences in 1916 to associate the broad community of science and technology with the Academy's purposes of furthering knowledge and advising the federal government. Functioning in accordance with general policies determined by the Academy, the Council has become the principal operating agency of both the National Academy of Sciences and the National Academy of Engineering in providing services to the government, the public, and the scientific and engineering communities. The Council is administered jointly by both Academies and the Institute of Medicine. Dr. Ralph J. Cicerone and Dr. Charles M. Vest are chair and vice chair, respectively, of the National Research Council.

www.national-academies.org

COMMITTEE ON THE INTERNATIONALIZATION OF THE CIVILIAN NUCLEAR FUEL CYCLE
NATIONAL RESEARCH COUNCIL OF THE NATIONAL ACADEMIES

JOHN F. AHEARNE, (*Chair)*, Sigma XI
ROBERT J. BUDNITZ, Lawrence Berkeley National Laboratory
MATTHEW BUNN, Harvard University
WILLIAM F. BURNS, Major General (USA, retired)
STEVE FETTER, University of Maryland
ROSE GOTTEMOELLER, Carnegie Moscow Center
MILTON LEVENSON, Bechtel International (retired)

COMMITTEE ON THE INTERNATIONALIZATION OF THE CIVILIAN NUCLEAR FUEL CYCLE
RUSSIAN ACADEMY OF SCIENCES

NIKOLAY P. LAVEROV, (*Chair)*, Russian Academy of Sciences
VALERY S. BEZZUBTSEV, Rostekhnadzor
ALEXANDER V. BYCHKOV, Research Institute of Atomic Reactors
VALENTIN B. IVANOV, Institute of Ore Deposits, Petrography, Mineralogy, and Geochemistry
BORIS F. MYASOEDOV, Russian Academy of Sciences
VLADISLAV A. PETROV, Institute of Ore Deposits, Petrography, Mineralogy, and Geochemistry
MIKHAIL I. SOLONIN, Technology and Innovation Center of the TVEL Corporation

National Research Council Staff

MICAH D. LOWENTHAL, Study Director, Nuclear and Radiation Studies Board
ANNE M. HARRINGTON, Director, Committee on International Security and Arms Control
RITA S. GUENTHER, Senior Program Associate, Committee on International Security and Arms Control

Russian Academy of Sciences Staff

YURI K. SHIYAN, Director, Office for North American Scientific Cooperation

PREFACE AND ACKNOWLEDGMENTS

The so-called nuclear renaissance has increased worldwide interest in nuclear power. This potential growth also has increased, in some quarters, concern that nonproliferation considerations are not being given sufficient attention. In particular, since the introduction of many new power reactors will lead to requiring an increase in uranium enrichment services to provide the reactor fuel, the proliferation risk of adding enrichment facilities in countries that do not have them now led to proposals to provide the needed fuel without requiring new indigenous enrichment facilities. Similar concerns exist for reprocessing facilities.

In 2006, International Atomic Energy Agency (IAEA) Director General Mohamed ElBaradei, Russian President Vladimir V. Putin, and U.S. President George W. Bush each announced plans to assure the provision of fuel to countries that want to develop nuclear power. The proposals were aimed at dissuading these countries from building uranium enrichment plants because such plants could be used to produce weapons-usable highly enriched uranium. In the spring of 2006, members of the Committees on International Security and Arms Control of the U.S. National Academy of Sciences (NAS) and the Russian Academy of Sciences (RAS), which have had a productive partnership for more than 25 years, met with each other, with senior officials in their respective governments, and with Director General ElBaradei to identify issues of national and international importance on which independent advice from the two academies would be useful.

With funding from the John D. and Catherine T. MacArthur Foundation and the Carnegie Corporation of New York, two committees with members appointed by the NAS and the RAS, working jointly, produced this report analyzing the proposals and options for future international nuclear fuel cycles, including the incentives that might be required for countries to accept fuel assurance guarantees and not develop enrichment or reprocessing facilities, as well as technical fuel-cycle issues. The statement of task for this study can be found in Appendix A. The task notes that this report is not intended to cover the policy and technical aspects of international fuel cycles comprehensively. Rather, the joint committees summarize key issues and analyses, offer some criteria for evaluating options, and make findings and recommendations to help the United States, the Russian Federation, and the international community reduce proliferation and other risks as nuclear power is used more widely.

This report is intended for all those who are concerned about the need for assuring fuel for new reactors and at the same time limiting the spread of nuclear weapons. This audience includes the United States and Russia, other nations that currently supply nuclear material and technology, many other countries contemplating starting or growing nuclear power programs, and the international organizations that support the safe, secure functioning of the international nuclear fuel cycle, most prominently the International Atomic Energy Agency.

Fuel assurance proposals have been discussed in conferences and journal articles. However, to receive input from the countries that might use the fuel assurance program, the joint committees held a workshop at the IAEA in April 2007, where people from eight countries presented their opinions or comments on the fuel assurance programs. While not officially representing their governments, these experts provided valuable insights into the issues that must be addressed for the fuel assurance programs to succeed. Appendix B of the report contains a summary of the workshop. The joint committees also addressed technologies being developed for new approaches to reprocessing (also called recycling and regeneration) and possible advanced reactors. While these discussions are necessarily limited due to the technologies being in the early stages of development or existing only as concepts, some advantages and disadvantages are discussed.

The joint committees addressed the different elements of the statement of task at different levels. Much of Part B of the task calls for comparisons of technologies in Russia with those envisaged in the United States. The Global Nuclear Energy Partnership (GNEP) comprises two initiatives from President George W. Bush. One is an international initiative beginning with an accord expressing the signatories' guiding principles for expansion of nuclear power. The other is a domestic nuclear energy and fuel cycle technology initiative with seven different goals. The international initiative has garnered dozens of partners. The domestic technology initiative has shifted its focus, emphasis, and timeline several times over the course of the study. These changes were significant, from switches among advanced fuel processing technologies that are mostly in the research phase and evolutionary commercial fuel-processing technologies to different fuels manufactured with as-yet-to-be-developed technologies. For these reasons, the joint committees were unable to compare the concrete Russian technological options with the multitude under consideration in GNEP. Because the Russian approaches have been developed more fully and in many cases the Russian government has selected particular approaches for deployment, these approaches are described in more detail in this report than the early-stage concepts being considered in the U.S. Technologies in related areas being pursued in other countries were beyond the joint committees' charge, and are considered only in passing here.

We wish to thank the IAEA, especially Director General Mohamed ElBaradei, Deputy Director General Yuri Sokolov, and Tariq Rauf for their support of our international workshop held in Vienna. We also thank Alan MacDonald of the IAEA for his substantial assistance in arranging the workshop. We thank the workshop attendees and the presenters at the joint committee meetings in the United States and Russia who provided us with their expert knowledge (see Appendix E).

We especially thank Yuri Shiyan of the RAS, Micah Lowenthal, NAS Study Director, and Rita Guenther of the NAS. Without the tireless work of these three individuals, the report would not have been completed.

This joint study addresses some of the serious international issues connecting the spread of nuclear power and nonproliferation concerns. The NAS and RAS have met and worked together for many decades on issues related to science and technology, including decades of dialogues and, more recently, joint studies on international security problems. We strongly believe that inhibiting the spread of nuclear weapons capabilities while promoting better access to safe, clean energy is in the interests of Russia, the United States, and the larger world community. It is precisely at times like these, then, that cooperation is needed between our scientific communities to help focus on those common interests and promote efforts toward common goals. The need for such cooperation grows under the conditions we see today.

John F. Ahearne	Nickolay P. Laverov
Committee Co-Chair	Committee Co-Chair
U.S. National Academies	Russian Academy of Sciences

ACKNOWLEDGMENTS

This report has been reviewed in draft form by individuals chosen for their diverse perspectives and technical expertise, in accordance with procedures approved by the National Academies' Report Review Committee. The purpose of this independent review is to provide candid and critical comments that will assist the institution in making its published report as sound as possible and to ensure that the report meets institutional standards for objectivity, evidence, and responsiveness to the study charge. The review comments and draft manuscript remain confidential to protect the integrity of the process.

We wish to thank the following individuals for their review of this report: R. Stephen Berry, The University of Chicago; Leonid Bolshov, Institute of Nuclear Safety; Douglas Chapin, MPR Associates, Inc.; Richard Garwin, IBM Watson Research Center (retired); David McAlees, Siemens Corporation (retired); Alan McDonald, IAEA; Leonam dos Santos Guimarães, Eletronuclear; Ashot Sarkisov, Institute of Nuclear Safety; Lawrence Scheinman, Center for Nonproliferation Studies; Mohamed Shaker, Egyptian Council for Foreign Affairs; Frank von Hippel, Princeton; Vassily Velichkin, Institute for Geology and Mineralogy; and Ray Wymer, ORNL (retired).

Although the reviewers listed above have provided many constructive comments and suggestions, they were not asked to endorse the conclusions or recommendations, nor did they see the final draft of the report before its release. The review of this report was overseen by Christopher Whipple, ENVIRON, and Harold Forsen, Bechtel Corporation. Appointed by the National Academies, they were responsible for making certain that an independent examination of this report was carried out in accordance with institutional procedures and that all review comments were carefully considered. Responsibility for the final content of this report rests entirely with the authoring joint committees and the institution.

CONTENTS

SUMMARY

Following the proposals for nuclear fuel assurance of International Atomic Energy Agency (IAEA) Director General Mohamed ElBaradei, former Russian President Vladimir V. Putin, and U.S. President George W. Bush, joint committees of the Russian Academy of Sciences (RAS) and the U.S. National Academies (NAS) were formed to address these and other fuel assurance concepts and their links to nonproliferation goals. The joint committees also addressed many technology issues relating to the fuel assurance concepts. This report provides background information and support for the following consensus findings and recommendations of the joint committees:

Finding 1a
By 2020, many countries that currently do not have a nuclear power plant are likely to initiate national programs for the construction of nuclear power stations.[1] These countries do not now have facilities for uranium enrichment for nuclear fuel production or spent nuclear fuel reprocessing.

Finding 1b
Uranium enrichment and spent fuel reprocessing are the key technologies that enable countries to produce direct-use materials for nuclear weapons.[2] The more countries to which either technology (enrichment or reprocessing) spreads, the greater the proliferation risks. Currently it appears that more countries that have not already deployed these technologies are interested in establishing uranium enrichment programs than in pursuing spent fuel reprocessing technologies, making the spread of enrichment technology a greater near-term concern for nuclear proliferation. But the intention to acquire spent nuclear fuel reprocessing capabilities was the main focus of proliferation concerns in the 1970s and could become so again.

Finding 1c
Requirements of the nuclear security environment, the difficulty of providing safeguards and security, and the demand for nuclear fuel cycle services change over time, and technology advances with time. Any approach for enhancing the nonproliferation features of international fuel cycles must be staged to respond to the nonproliferation needs of the time period. Today this suggests a focus on convincing countries that they do not need to establish their own enrichment facilities, which has motivated efforts by several countries and international organizations to address the enrichment issue. Similar efforts are needed to convince countries that they do not need their own reprocessing facilities. Also needed are strengthened efforts to prevent the spread of these technologies through illicit or inadequately regulated exports and black-market nuclear networks, and improved safeguards for both uranium enrichment and spent

[1] Until and unless construction begins, estimates of nuclear growth are based upon expressions of interest and should be considered as having substantial uncertainty.
[2] The main nuclear weapons materials are highly enriched uranium, obtained by enriching naturally occurring uranium, and plutonium, primarily obtained by reprocessing irradiated reactor fuel.

fuel reprocessing facilities, designed both to increase international confidence that significant diversions from declared facilities would be detected and to strengthen the ability to provide timely warning concerning covert facilities and activities.

Recommendation 1a
The countries that currently provide nuclear fuel services should redouble efforts, with other countries and the IAEA, to establish mechanisms for increasing reliability of supply of nuclear fuel, so that countries that do not now have enrichment technology would have reduced incentives to build their own uranium enrichment facilities.

Recommendation 1b
The international community should help countries provide adequate capacity for safely storing spent fuel (on their own territory or elsewhere), or reliable reprocessing services from existing providers, to reduce countries' incentives to establish their own reprocessing facilities. Separated plutonium or fabricated plutonium fuel should not be sent to countries that have not previously received such material and do not have reprocessing capabilities. The spread of separated plutonium to additional countries poses many of the same proliferation risks posed by the spread of reprocessing capabilities.

Recommendation 1c
For similar reasons the United States and other nations should reduce and seek to minimize commerce in and the transfer of highly enriched uranium (which poses proliferation risks) except if sealed in a reactor core.

Second-level findings:

a. To ensure a reliable supply of nuclear fuel, a country needs reliable fuel fabrication services as much as it needs reliable sources of uranium and enrichment services.
b. To assist in the international fuel assurance programs, it would be helpful if nations with fuel fabrication facilities made those available.
c. Fuel fabrication technology for uranium oxide fuel with low-enriched uranium is not sensitive from a proliferation perspective. Hence, if countries choose to establish their own fabrication capabilities to produce fuel assemblies for their own nuclear power stations, without establishing uranium enrichment or spent fuel reprocessing capabilities—as South Korea has done, for example—this should not pose significant international concerns.

Finding 2
Several messages are clear from the NAS-RAS workshop and other recent discussions in Vienna about assurance of supply:

a. Few countries have declared a willingness to forgo forever a right to develop their own uranium enrichment or spent fuel reprocessing nuclear technology in the future.[3]

[3] The charter of the International Uranium Enrichment Center in Angarsk, Russia, requires members other than the host country to commit to not develop their own uranium enrichment capabilities. As of June 2008, Kazakhstan and Armenia have made that commitment and become members.

Some countries have expressed adamant opposition to requiring a country to forgo the development of its own enrichment and reprocessing technologies as a condition of assurance of supply of nuclear fuel or low-enriched uranium.

b. To be successful, uranium enrichment, fuel assembly production for nuclear power stations, and spent fuel storage/reprocessing technologies continue to operate in the international market.
c. No single mechanism or strategy for assurance of nuclear fuel supply is likely to address every country's legitimate needs and desires. Each country's or region's needs and requirements may be different.
d. New mechanisms for assured nuclear fuel supply may only modestly change countries' incentives to establish enrichment facilities, as the existing international market provides strong assurance of supply, and countries have a variety of other reasons for establishing their own enrichment plants, including a desire to participate in the profits of enrichment, national pride, and a desire to establish a nuclear weapons option for the future.

Recommendation 2a
The governments of the United States and Russia should continue to support a broad menu of approaches to increasing assurance of nuclear fuel supply.

An array of mechanisms for assurance of nuclear fuel supply has been proposed, from diversified long-term contracts through the existing market, enrichment bonds,[4] and international fuel centers to creating a virtual or actual fuel bank. Some of these are already in place. The Russian and U.S. governments should support a broad menu of these approaches, ensuring that these do not undermine each other.

Recommendation 2b
The governments of the United States and Russia should seek to establish additional benefits and incentives for countries that choose not to establish their own uranium enrichment and spent fuel reprocessing facilities. Possibilities could include assistance in establishing the necessary infrastructure for safe and secure use of nuclear energy.

Recommendation 2c
To support nonproliferation goals, the nations that currently supply nuclear fuel should work expeditiously with other countries and the IAEA to make assured fuel supplies available before there is a major commitment to new nuclear power plants by countries that do not have them today.

Finding 3a
It is feasible to establish a multinational center to provide enrichment services without sharing enrichment technology for countries willing to refrain from developing their own enrichment

[4] Enrichment bonds: A guarantee by a state that supplies enrichment services that enrichment providers will not be prevented from supplying the recipient state with uranium enrichment services if the guarantee is invoked (adapted from a proposal by the United Kingdom).

facility as long as they participate in the center.[5] The International Uranium Enrichment Center (IUEC) in Angarsk, Russia, is one such center. There have been proposals to establish centers under international organizations, although their feasibility has yet to be established. An international dialog, in which concerned countries evaluate the pros and cons of supplementing multinational centers with a center under international control, is needed. Two European multinational consortia have provided enrichment services for two decades: Eurodif, like the IUEC, does not share its technology among its members, but participants need not forgo development of enrichment technology as a condition of participation. Urenco has only three partners, all of which have access to its technology.

Finding 3b
If global usage of nuclear energy increases, it may become increasingly difficult to maintain a system in which nationally controlled facilities in only a few countries provide all enrichment and reprocessing services, as desirable as that might be from a nonproliferation perspective. Offering the opportunity to profit from these technologies may reduce the likelihood that countries would perceive efforts to inhibit expansion of access to the technology as unfair.

Recommendation 3
Over time, Russia, the United States, and other nations should work to create a global system featuring a small number of centers for the sensitive steps of the fuel cycle (especially enrichment and spent fuel management, possibly including storage, reprocessing, or disposal), owned, operated, and controlled by consortia of states or international organizations (but without spreading the relevant technologies beyond existing technology holders). Such a global system, offering many countries the opportunity to participate and share in the profits, would provide a somewhat more equitable and sustainable long-term basis for limiting enrichment and reprocessing facilities to a small number of countries. There has been some criticism that the proposed mechanisms are unfair. The preliminary arrangements should be improved over time.

Finding 4
As use of nuclear power grows, there is a need worldwide for well-educated personnel to support the whole nuclear fuel cycle.

Recommendation 4
Countries with large nuclear power programs, such as the United States and Russia, should encourage young people to enter nuclear engineering and related fields and programs that give the breadth of perspective needed.

Finding 5
Arrangements that would provide assured return of spent nuclear fuel could provide a much more powerful incentive for countries to rely on international nuclear fuel supply than would assured supply of fresh fuel, because assured take-back could mean that countries would not need to incur the cost and uncertainty of trying to establish their own repositories for spent

---

[5] By a *multinational* center, the joint committees mean a facility whose ownership and management involves an arrangement among several countries. Eurodif, Urenco, and the International Uranium Enrichment Center at Angarsk are examples. By an *international* facility, the joint committees mean a facility whose ownership and management are centered in a fully international organization such as the IAEA.

nuclear fuel or nuclear waste. Further, it would reduce the number of countries where plutonium-bearing material is stored around the world. Fuel leasing, reactor leasing, and similar approaches could have this benefit, if managed appropriately. For many countries, however, the political barriers to taking back other countries' spent nuclear fuel or nuclear waste are substantial.

Recommendation 5

The United States, Russia, and other suppliers should increase their emphasis on establishing mechanisms for assured fuel-leasing or reactor-leasing services,[6] including take-back of all irradiated fuel. Russia already has legislation and arrangements in place to offer fuel leasing and has such a contract in place with Iran. In both international fuel supply approaches and in take-back of spent fuel, Russia is farther along in offering services to other countries. The United States and Russia should work together on cooperative approaches that would make it possible to enter into fuel-leasing arrangements in which they would guarantee to supply, and to take back, fuel for the lifetime of reactors built in "newcomer" states, with the fuel taken back to Russia for now, or to the United States, as well, if circumstances someday make that possible.

Finding 6

A hidden danger of creating such centers is the potential for leakage of sensitive technology. The most damaging leakage of sensitive technology occurred when A. Q. Khan, working as a contractor for Urenco, was able to acquire enough information and contacts to build the supply line for Pakistan's nuclear weapons program. Khan went on to form a supply network that fed into weapons programs in Libya, North Korea, and Iran. An event like this puts the nonproliferation regime in great danger.

Recommendation 6a

The United States and Russia should work diligently with other nations to ensure that all efforts to establish international centers for enrichment, reprocessing, or other sensitive activities include specific, stringent plans to prevent leakage of sensitive information and technology. Plants with staff from countries that do not have technology of the type used at that plant should maintain the sensitive technology in "black boxes" so that the international staff does not have access to the technologies themselves. Plans to prevent technology leakage should be subject to review by a small group of international experts familiar with such technology controls before the centers are established.

Recommendation 6b

Russia the United States and other countries working to develop centers should have criteria for participation. Two major criteria for participation by countries beyond the technology holders who provide the technology for the center should be that they not have or be developing an enrichment facility, and that they should be in compliance with IAEA safeguards and nonproliferation obligations.

---

[6] Today the only discussions of reactor leasing are those on the floating power plants being built by Russia and the nuclear battery being proposed by Toshiba. There will be many legal issues to work out in both cases.

Finding 7
Safeguard arrangements, fuel transfer processes, and return of spent fuel provisions are only a few of the complex legal issues that must be resolved if fuel assurance, fuel take-back, and multinational or international fuel center programs are to be effective.

Recommendation 7
The IAEA should lead an international effort to identify these legal questions and options to be considered. The IAEA should also convene countries to reach agreement on preferred solutions.

Finding 8
Both Russia and the United States are working on new technologies for processing spent fuel, intended to reduce the economic costs and proliferation risks of traditional reprocessing approaches and improve waste management. The technologies being proposed would still pose significant proliferation concerns if deployed in countries that did not previously have reprocessing capabilities. The new technologies under development will take significant time before being ready for demonstration at commercial scale.

Recommendation 8
Developers of nuclear fuel cycle technologies should assess the technologies' proliferation risks and projected economic costs and benefits as critical elements of design.

Finding 9a
In most cases, reprocessing is not economic under current conditions. When the world's economically recoverable uranium resources diminish compared to demand or there is widespread deployment of fast reactors, then reprocessing may become economically attractive.

Finding 9b
Excess stocks of plutonium separated from spent fuel, beyond plutonium that would be needed for making MOX fuel for use in the near term, pose security risks.

Recommendation 9
States should end the accumulation of stockpiles of plutonium separated from spent fuel as soon as practicable, and begin to reduce existing stocks. Spent fuel should only be reprocessed when its constituents are needed for fuel, or when reprocessing is necessary for safety reasons.

Finding 10
Many of the technologies for improved nuclear fuel cycles are not areas that will advance without directed research specifically focused on the nuclear fuel cycle; advances in other areas of science and engineering will help, but are not sufficiently linked to nuclear fuel cycles to solve the technical challenges described here, by themselves. Research is needed in the areas of processing of irradiated nuclear fuel and nuclear fuel design (beyond the incremental improvements in uranium oxide fuel for light water reactors), as well as in improved approaches to disposal of wastes or spent fuel, and reduced-cost recovery of uranium from low-grade sources. Additional research and development is also needed to develop advanced safeguards and security technologies that can provide increased capabilities to detect covert nuclear

facilities; highly accurate near-real-time monitoring of material flows in bulk processing plants with reduced intrusiveness, increasing confidence that any diversion would be detected; low-cost real-time monitoring that would set off an immediate alarm if stored nuclear material were tampered with or removed; effective protection against sophisticated outsider and insider theft and sabotage threats at reduced cost; and design of facilities to simplify and increase the effectiveness of safeguards.

Recommendation 10
The U.S., Russian, and other governments should take the lead in a cooperative international effort to make additional research and development investment in advanced safeguards and security technologies. A focused effort should be made to make the results of this research and development available to the international community to ensure that new facilities are more secure and readily safeguarded. The international community also should adopt the philosophy of designing high levels of security and safeguards into new nuclear systems and facilities from the outset, including both the inherent technical characteristics of the process and the institutional measures to be taken.

Finding 11
It is not possible today to construct an entire, operational international fuel cycle program.[7] Such a program will have to be built incrementally. However, elements of that program currently exist and the groundwork for other elements has been laid.

Recommendation 11
The U.S., Russian, and other governments should

- continue to invest in research and development on advanced approaches to once-through and closed fuel cycles that offer the potential to improve proliferation resistance, safety, security, economics, resource utilization, and waste management
- utilize a systems approach to developing and assessing these technologies, with clear objectives and technically justifiable criteria for decision making. Use systems analysis to identify potentially promising approaches before proceeding to build pilot or larger facilities.
- take all relevant proliferation risks into account when assessing proliferation resistance, including how the availability of the materials, facilities, and expertise associated with a particular fuel cycle approach would affect the time, cost, uncertainty, and detectability of a nuclear weapons program

The implementation of those elements that are feasible today, for example, assurance of fuel supply, should not be delayed while other options are being refined or explored both institutionally and technically.

[7] One run internationally and including all elements of the fuel cycle.

Finding 12
The United States and the Russian Federation have signed an agreement on peaceful nuclear cooperation, but it must still be allowed to come into force. The lack of a U.S.-Russian agreement in force is interfering with joint efforts to reduce proliferation. The expanded cooperation in nuclear energy research and development and commercial implementation that such a bilateral cooperation could make possible could serve both countries' interests in expanding the use of nuclear energy while meeting safety, security, and nonproliferation objectives. Article 2 of the signed agreement lists possible areas of cooperation, including, among other areas, scientific research and development on nuclear power reactors and their fuel cycles; nuclear fuel cycle services; radioactive waste handling; and nuclear safety, regulation, nonproliferation, and safeguards.

The joint committees recognize that it is unlikely that the U.S. government will bring the agreement into force in an environment of worsening relations between the United States and Russia. It is the joint committees' hope that current disagreements that have recently emerged will not interfere with the United States and Russia working together toward their common goal of inhibiting nuclear weapons proliferation as nuclear energy use grows across the world.

CHAPTER 1

INTRODUCTION

After the introduction of full-scale nuclear power plants in the 1960s, many nuclear-generating stations were built and complemented by the construction of some fuel cycle facilities to support those stations. Growth of nuclear power slowed in most countries in the 1980s and 1990s. There is now substantial worldwide interest in building new nuclear power plants. This interest is evident not only in countries that led the world in the development of nuclear power—Canada, France, Russia, the United Kingdom, and the United States—but also in developing countries with large economies, such as China and India, and small economies, such as Belarus and Egypt. The current increased interest has been called a nuclear renaissance, because after years of relatively slow worldwide growth, many countries that do not have a nuclear power plant are considering building one; and many nations that already have one or more nuclear power plants are considering adding more nuclear power plants and expanding their nuclear enterprises with fuel fabrication, uranium enrichment, and (in at least one case) spent fuel reprocessing facilities to serve an expanded fleet of nuclear power plants.

According to the director general of the International Atomic Energy Agency (IAEA), in 2007, the IAEA was assisting with energy planning studies for 29 nations that are exploring nuclear energy as a potential option (ElBaradei, 2007).

> Countries such as Algeria, Belarus, Egypt, Indonesia, the Islamic Republic of Iran, Jordan, the Libyan Arab Jamahiriya, Nigeria, Thailand, Turkey, Vietnam and Yemen are among those considering or moving forward with the infrastructure needed to introduce nuclear power programmes. And many others, such as Argentina, Bulgaria, China, Finland, France, India, Japan, the Republic of Korea, Pakistan, South Africa, the Russian Federation and the United States of America are working to add new reactors to their existing programmes.

This potential large expansion of nuclear power carries with it a growing concern about proliferation of nuclear materials and the capability to manufacture nuclear weapons to other countries. The same technologies that are needed to enrich uranium to make reactor fuel and to separate plutonium from spent fuel to be used in fresh reactor fuel can be used to produce the fissile material needed for nuclear weapons. So if countries pursuing a nuclear energy strategy develop domestic enrichment or reprocessing technologies, or both, to ensure a supply of civilian nuclear fuel or manage their spent fuel, they will also acquire the means to create material that is directly usable in nuclear weapons.[1]

The director general of the IAEA, former President Vladimir V. Putin of Russia, President George W. Bush of the United States, and at least six other world leaders and organizations have proposed approaches to multinational or international fuel cycle facilities or fuel supply assurances. The goal is to reduce the likelihood of, or inhibit, the spread of enrichment and reprocessing to other countries by eliminating one motive for acquiring enrichment and reprocessing technologies.[2] With funding from the John D. and Catherine T. MacArthur Foundation and the Carnegie Corporation of New York, the U.S. National Academy of Sciences (NAS) and the Russian Academy of Sciences (RAS) assembled two committees of experts to carry out this joint consensus study on how to evaluate schemes for structured internationalization of parts of the nuclear fuel cycle, including both institutional arrangements and technical options. The statement of task can be found in Appendix A, and brief biographical sketches of the joint committee members are in Appendix F.

Five key motivations have spurred proposals for multinational or international fuel cycle approaches:

- **Assured fuel supply or spent fuel management.** Countries may feel more assured that they will always have reliable fuel supply for their reactors (and therefore have less incentive to build their own enrichment plants) if, for example, they are participants and part owners of a multinational enrichment plant, or if there are international mechanisms in place to provide backup supplies if a supply interruption occurs. International arrangements that would allow countries to send their spent fuel away after it was used could substantially reduce countries' incentives to invest in reprocessing plants—and fuel-leasing arrangements, in which fuel would be supplied with a promise to remove the spent fuel later,[3] could create particularly strong incentives for states to rely on an international fuel supply rather than having to invest in both their own fresh fuel facilities and spent fuel management facilities. A variant on the fuel-leasing idea is reactor leasing, where a sealed reactor with a core of long-life fuel is leased and then returned to the vendor unopened. (These arrangements are discussed further later in the report.)

- **Opportunities to participate in fuel cycle profits and management.** If countries can have a share in the profits from enrichment or reprocessing, and take part in the management of an enrichment or reprocessing enterprise, by taking part in a multinational facility in another state, this may reduce their incentive to invest in an enrichment and reprocessing plant of their own. Kazakhstan, for example, after joining Russia's International Uranium Enrichment Center at Angarsk, indicated that it was no longer interested in building its own enrichment plant.

[1] The IAEA defines "*unirradiated direct use material*" as nuclear material that can be used for the manufacture of nuclear explosive devices without transmutation or further enrichment, including unirradiated plutonium containing less than 80 percent Pu-238, uranium enriched to 20 percent or higher in the isotope U-235, and U-233.

[2] See IAEA, 2005b, for a description of the context and options as laid out by IAEA.

[3] The United States will have difficulty in convincing nations to accept its word. Examples such as the supercollider, the international space station, and the International Thermonuclear Experimental Reactor (ITER) project indicate that the United States can be an unreliable partner. The United States must overcome this attitude for it to become a trusted participant in a fuel assurance program.

- **Reduced proliferation risks from the plants that are built.** If an enrichment or reprocessing plant were owned by several countries or by an international organization, and operated by an international staff, this could provide both greater international transparency to detect any effort to use the plant for military purposes and a higher political barrier to doing so than would be present in a purely nationally owned and staffed facility. The daily interactions between the international staff and host-country experts might also make it more difficult to use those experts to establish a covert facility without detection. On the other hand, such approaches would have to be carefully structured to avoid unduly spreading knowledge of how to build and operate enrichment or reprocessing facilities: Sensitive fuel cycle facilities with staff from many countries could increase the risk of technology leakage, if effective controls on sensitive technologies were not put in place.

- **Pooling resources.** States may choose to pursue multinational approaches to bring the resources of several countries to bear on the problem. The German-Dutch-British Urenco consortium, for example, appears to have been motivated in large part to reduce the burdens that any one of these countries would have faced in developing an enrichment plant on their own. Similarly, there are a variety of proposals for international nuclear waste disposal facilities, to avoid the need for scores of countries to each have their own nuclear waste repository.

- **Removing materials that pose proliferation risks.** Finally, there have been several cases in recent years where nuclear material in a particular location was judged to pose a significant proliferation risk, and was removed. Discussions with North Korea about shipping its plutonium elsewhere are ongoing. International spent fuel or nuclear waste management facilities might provide ready-made institutions that could receive high-risk materials, making such removals of high-risk materials easier to carry out; offers to remove spent fuel for storage or processing in other countries could avoid accumulating large stocks of plutonium-bearing spent fuel in many countries as nuclear energy expands and spreads in the future.

In essence, the joint committees asked, acknowledging that countries must be able to fuel their nuclear plants reliably and manage their spent fuel to take part in this renaissance, How then can the expected international expansion of the use of nuclear power proceed without spreading the capability to make nuclear weapons? This report contributes an assessment of this issue and provides some possible approaches to resolve it. Any of these approaches, however, would be only one part of a broader strategy to reduce the risk of nuclear proliferation.

Addressing this question of how to expand nuclear power without spreading nuclear weapons capabilities requires understanding of the countries' needs and desires for nuclear power, what factors would prevent a nation from developing or acquiring nuclear weapons capabilities, as well as the factors that increase or undermine countries' trust in systems that promote nuclear nonproliferation. Each of these issues is discussed briefly here and in more detail below.

To learn about a spectrum of countries' needs and interests, the joint committees monitored developments related to nuclear fuel cycles in their home countries, attended international events focused on these topics, and surveyed the literature. In addition, the joint committees convened an international workshop with the assistance of the IAEA (more details on the workshop, including the list of participants and a summary of the discussion are in Appendix B). While the workshop was small, it was diverse. Yet despite the differences among the workshop participants' home countries, some common messages emerged, including (a) their imperative to maintain sovereign rights to develop peaceful technology and the corresponding rejection of any idea that they would sign agreements never to enrich uranium or reprocess spent nuclear fuel (that is, forgo those technologies); (b) the desire to protect the functioning market for uranium, uranium enrichment, and fuel fabrication; and (c) the observation that take-back of spent nuclear fuel by the supplying country (or even another country) would be a larger incentive than assurance of nuclear fuel supply.

WHY IS THERE INTEREST NOW IN NUCLEAR POWER?

Several factors are driving the increased interest in nuclear power: growth in energy demand; increased costs and projected limited supplies of fossil fuels; safer, less costly,[4] and more efficient nuclear power plants, and better management experience with existing plants; and concerns about global climate change. As is explained extensively later in this report, each nation has its own set of interests and needs for its energy sector. The United States, which has the largest nuclear energy enterprise in the world but has built no nuclear power plants ordered after the early 1970s, illustrates several of the changes that have led to the renewed interest.

In the United States, some owners or prospective owners of nuclear power plants in the 1970s and 1980s incurred substantial financial losses. The causes were many and are still debated, but the principal reasons were a sharp drop in the growth of electricity demand following the oil embargoes in the mid-1970s, and mismanagement of construction of these capital-intensive facilities at a time when interest rates for financing were high. In the United States, once these power plants came online they often operated with relatively low capacity factors (they did not produce electricity for all the time that, in principle, they could have) and with the need to recover the large capital costs, nuclear plants were not cost competitive with other base-load power plants, such as coal, hydropower, or even natural gas. Reactor accidents in the United States and the Soviet Union (Three Mile Island in 1979 and Chernobyl in 1986, respectively) led many people to conclude that nuclear power is unacceptably unsafe. Some people have viewed nuclear power as environmentally unfriendly because of concerns about radioactive waste, particularly the persistent difficulties in approving a repository location in which to bury and hence dispose of the spent fuel from reactors. All of these factors, to varying degrees, resulted in much slower growth of nuclear power than was anticipated in the 1960s and early 1970s.

Several of these factors began to change in the 1990s. Shorter times to construct a nuclear power plant and bring it online have been demonstrated in France, Japan, and the Republic of Korea. Capacity factors of nuclear power plants have risen, and many plants are operating near their theoretical maximum capacity factor (over 90 percent on average, compared with about 60 percent in some earlier years). Also, the 40-year operating licenses on many

[4] New designs have much less cabling, fewer pumps, and a reduced number of safety systems, all related to passive rather than active shutdown systems. Coupled with modular construction, these changes should make the new plants less costly than if a current operating design were to be built today.

existing nuclear power plants are being extended by 20 years. The ability to reap 20 more years of electricity generation from power plants, especially existing plants that have already defrayed their capital costs, makes them more attractive financially. Costs of natural gas, the main competitor in the United States for new plants, have risen. Recognition that the Three Mile Island accident was not the nuclear disaster some feared, and the ongoing good safety record of nuclear power over the last 20 years, have mitigated some of the worries about safety of nuclear power plants. Growing concern about emissions of greenhouse gases that contribute to global climate change has improved the environmental credentials of nuclear power plants, which over their life cycle can have very small greenhouse gas emissions to the atmosphere[5] in contrast to the large amounts of carbon dioxide from coal plants, the largest source of U.S. electricity generation. Nuclear power is increasingly seen as an environmentally responsible alternative for meeting expanding demand for base-load power.[6] That demand is increasing without showing signs of stopping.

Experience in Russia, another large nuclear enterprise, is also instructive. Before its dissolution, the former Soviet Union prepared extensive plans for developing the nuclear power industry. Those plans included light-water-moderated reactors generating electrical power of 1,000 or 1,500 MW, as well as up to 20 units containing fast breeders of BN-600 type. However, the Chernobyl nuclear plant accident and the Soviet Union's collapse followed by the financial crisis in the 1990s put an end to such plans. The Chernobyl accident and low prices for hydrocarbons (oil and gas) were unquestionably the main cause of a drastic change in the Russian public's attitude toward nuclear power in the 1990s. Only three nuclear power plants (VVER-1000 type) were commissioned, and the construction of another plant (the BN-800 fast-breeder reactor) saw little progress between 1990 and 2005 because of insufficient funding.

The picture for the future is completely different in Russia today. The Russian government and parliament have adopted a decision in the form of the Federal Special-purpose Program providing for (a) construction of about 20 VVER-1000-type nuclear power plants to replace the legacy nuclear plants, and (b) completion of construction of the BN-800 fast-breeder reactor as a transition to nuclear plants with a new generation of fast breeder reactors to be used in a closed fuel cycle. Most people in Russia now do not object to further development of nuclear power (*Nuclear Power Today,* 2005).

At present, 31 power-generating units with an aggregate capacity of 23.2 gigawatts-electric (GWe) are operating in Russia. Nuclear power stations produce approximately 17 percent of the total electrical energy yield.

In 2006, the Russian government adopted a new targeted program, "Russian Nuclear Power Industry Sector Development in 2007-2010 and Until 2015," which would receive

---

[5] Estimates of greenhouse gas emissions from nuclear power vary dramatically, ranging from 3.5 to 100 g (carbon dioxide equivalent per kilowatt-hour electric: CO_2-eq./kWh electric). Enrichment of uranium for fuel is currently the biggest contributor to the emissions within the nuclear fuel cycle and varies depending on the technology employed (gaseous diffusion consumes 24 to 60 times as much energy as gas centrifuge enrichment) and the source of electricity used for enrichment (the country's fuel mix). Fthenakis and Kim, 2007, found a range of 17 to 39 g CO_2-eq./kWh for solar electric in the southwestern United States (the region of that country that is most amenable to solar power), and 16 to 55 g CO_2-eq./kWh for nuclear energy in the United States. High-efficiency coal-fired power plants emit about 1.05 kg CO_2-eq./kWh electric.

[6] Hydropower, the other major base-load power source with low carbon emissions, is inherently limited by the availability of suitable sites and is increasingly viewed as destructive to river ecosystems and a potential safety hazard.

1,471.4 billion rubles[7] in appropriations, including 674.8 billion rubles from the federal budget (see Appendix C). The program is to be implemented in two phases over a period of nine years. The main objectives of the program include an increased pace of development of the Russian nuclear power industry by means of putting new standard power-generating units into operation at the nuclear power stations with a total nominal electrical capacity of 2 GW per year or more with a 5-year construction cycle. By the end of the program's term, 10 new power-generating units with a total nominal electrical capacity of 9.8 GW or more will be put into operation, and another 10 power-generating units will be in various stages of construction. A fast-reactor power-generating unit with a capacity of 800 MW will be put into operation, which is planned to be used as a testing facility for the closed nuclear fuel cycle technology including the recycling of uranium-235 (U-235) and plutonium-239 (Pu-239) separated in the processing of spent nuclear fuel. Further, Russia plans to increase the share of nuclear power stations in Russia's total electricity output to approximately 25 percent from 15.9 percent in 2006 (IAEA, 2007b).

According to IAEA analysis (IAEA, 2007b), the contribution of nuclear energy to electricity generation worldwide is expected to grow at an annual rate of between 0.9 and 2.8 percent. This would bring the per capita electricity demand from the 2006 level of 2.7 MWe-h/yr to a world average of 2.9 to 3.6 MWe-h/yr by 2020 and 3.2 to 4.8 MWe-h/yr by 2030. The same analysis projects that the population will grow from 6.5 billion people to 7.5 billion by 2020 and to 8.1 billion people by 2030, meaning that total electricity demands might more than double in approximately the next 20 years (IAEA, 2007b), rising from 17,550 TWe-h/yr to at least 22,000 TWe-h/yr and as much as 27,000 TWe-h/yr by 2020, and at least 26,000 TWe-h/yr and as much as 39,000 TWe-h/yr by 2030. One large but typical single-reactor nuclear power plant (1.25 GWe operating at just over 90 percent capacity factor) can produce about 10 TWe-h/yr. The IAEA estimates that nuclear power-generating capacity will increase from 369.7 GWe in 2006 to 447-691 GWe in 2030 (IAEA, 2007b).[8] Analysis by the Organisation for Economic Co-operation and Development (OECD) (OECD 2006; IAEA 2006b), demonstrates that such rapid growth of nuclear energy will be accompanied by a substantial increase in the demand for natural uranium. The recent edition of the OECD/IAEA uranium resource summary (the *Red Book*) states that the currently identified resources are adequate to meet the forecasted expansion from 372 GWe in 2007 to up to 663 GWe in 2030 (OECD/IAEA, 2008; Schneider and Sailor, 2008).[9]

To satisfy the demand for uranium through 2020, it is necessary to significantly intensify geological efforts, and by 2030 to bring into operation tens of new mines, which can exceed the current excavation levels two to three times. Solving this problem is entirely possible. Moreover, natural uranium is not directly related to the problem of the proliferation of weapons-grade nuclear material.

However, a substantial increase in demand for fresh nuclear fuel, growth in uranium enrichment and the generation of still greater amounts of spent nuclear fuel and possible reprocessing (a change of approximately 1,300 to 3,400 metric tons of fuel per year by 2020,[10] if

[7] At the time this report was issued, 1 U.S. dollar was worth approximately 20 rubles.

[8] This does not account for replacing reactors that might retire during that period.

[9] There are good reasons to believe that IAEA estimates may prove to be conservative as high prices motivate more exploration; in the case of most other minerals, real prices have fallen with time and quantity extracted, as technological improvements outpaced the depletion of the lowest-cost ores.

[10] These numbers assume burn-up in the range of 50 to 60 GW-days/MTHM. One MTHM is a metric ton of heavy metal (for example, uranium or plutonium) initially in the reactor fuel.

these were all light-water reactors), significantly intensifies the possible unauthorized proliferation of sensitive nuclear technologies and weapons-grade materials.

Current consumption and anticipated growth in demand are not distributed equally across nations. Africa is expected to continue to consume less electricity per capita than any other continent (0.7 to 1.1 MWe-h/yr) and North America is expected to consume more than others (14.8 to18 MWe-h/yr) (IAEA, 2007b), but overall consumption on these continents is not changing quickly. By contrast, East Asia, the Middle East, and South Asia saw 5 percent annual increases in electricity consumption between 1996 and 2006.

To continue that growth, nations are looking to develop electrical-generating capacity from nearly every resource available, including nuclear power. A recent U.S. review listed the following countries as giving serious consideration to nuclear power in the next 10 years: Azerbaijan, Belarus, Egypt, Estonia, Indonesia, Kazakhstan, Latvia, Norway, Poland, Turkey, and Vietnam. The same review listed the following countries with longer term plans under way: Algeria, Australia, Bahrain, Chile, Georgia, Ghana, Jordan, Kuwait, Libya, Malaysia, Morocco, Namibia, Nigeria, Oman, Qatar, Saudi Arabia, Syria, the United Arab Emirates, Venezuela, and Yemen (ISAB, 2008). The reasons for anticipated growth of nuclear power in a small number of other countries are described in Appendix B, the summary of the NAS-RAS international workshop held in Vienna in 2007.

Finding 1a

By 2020, many countries that currently do not have a nuclear power plant are likely to initiate national programs for the construction of nuclear power stations.[11] These countries do not now have facilities for uranium enrichment for nuclear fuel production or spent nuclear fuel reprocessing.

THE PROLIFERATION PROBLEM IN MORE DETAIL

A nation seeking to acquire nuclear weapons needs direct-use nuclear material and the knowledge and means to make that material into a weapon. It is generally assumed that the knowledge required to make at least a rudimentary nuclear weapon is available or fairly readily acquired. The difficulty of acquiring the direct-use nuclear material is the greatest technical barrier for a nation seeking to develop its own nuclear weapon, though political, diplomatic, and military pressure, either external or internal, may also lead countries to slow or reverse their nuclear programs.

Uranium enrichment facilities and nuclear fuel reprocessing facilities used for peaceful nuclear energy objectives (serving civilian nuclear power plants) can also be used to create direct-use nuclear material for weapons.

Uranium-235 (U-235) is the easiest material from which to fabricate a crude nuclear explosive device, although substantially more material is needed to construct an efficient nuclear explosive device using uranium-235 compared to plutonium-239. Uranium-235 occurs naturally in very low concentration (0.7 percent) in natural uranium mined from the earth. Another isotope, uranium-238 (U-238), constitutes nearly all of the rest of the natural uranium. Natural uranium can be used to fuel a nuclear power reactor (CANDU [Canadian deuterium-uranium]

---

[11] Until and unless construction begins, estimates of nuclear growth are based upon expressions of interest and should be considered as having substantial uncertainty.

reactors have done so), but higher concentrations of uranium-235 in nuclear fuel enable the fuel to sustain a fission chain reaction more readily in a relatively compact core using ordinary water (light water, which absorbs some neutrons) as the moderator. The process that raises the concentration of a particular constituent of a feed material in the product stream is called enrichment. Most nuclear power reactors in the world (called light-water reactors) require fuel enriched in uranium-235 to about 3-5 percent. Nuclear power stations with so-called fast reactors require higher enrichment—approximately 15 percent or higher, and more typically using around 20 percent.

Uranium enriched below 20 percent is called low-enriched uranium (LEU), and uranium enriched to 20 percent and higher is called highly enriched uranium (HEU). HEU can be used to construct nuclear weapons. Although a weapon with 20 percent enrichment is theoretically possible, the mass of uranium required makes such a weapon impractical (see Figure 1-1); higher concentrations of uranium-235 are more effective for use in weapons. But even possession of LEU is of some concern, particularly when it is coupled with further LEU enrichment capabilities, as explained below.

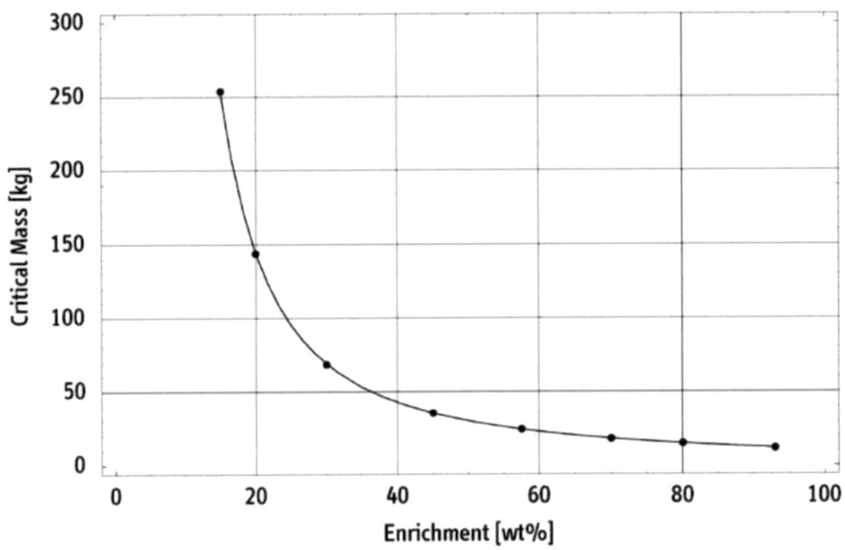

Figure 1-1 Critical mass of a metal uranium sphere with a 10-cm beryllium reflector as a function of the uranium-235 enrichment (weight percent, wt%). SOURCE: Glaser (2006).

There is no fundamental technological difference between a uranium enrichment facility used for civilian nuclear fuel and one used to produce HEU for weapons. Enrichment facilities are expensive and technologically challenging to develop and operate. Some enrichment facilities are also difficult to detect via satellite imagery, emissions, or any other observations, and this is only getting more difficult as enrichment technology improves.[12] LEU is not inherently a proliferation concern, but the work in terms of time and energy usage required to enrich a given amount of natural uranium to 3 percent uranium-235 is more (depending on

[12] The 1991 discovery of the Iraqi electromagnetic isotope separation (calutron) program came from information provided by a defector. Centrifuge facilities are the main concern now. Laser isotope separation, too, would be difficult to detect.

specifications of the enrichment, substantially more) than the work required to raise that 3 percent-enriched uranium to 90 percent uranium-235, which is not just weapons usable, but weapons grade (that is, what nuclear weapons states use in their weapons). The time and energy required to enrich uranium from 5 percent to 90 percent is lower still.[13] Thus, if a campaign to create direct-use material were to start with 5 percent enriched uranium as its feed, the lead time for acquiring significant quantities of HEU would be about one-fourth the time required if using natural uranium in the same enrichment facility. Such scenarios are called breakout scenarios.

It is more challenging to design a nuclear weapon with plutonium-239, but there are advantages to such weapons, such as the ability to mount them more easily on missiles because they can be more compact. Plutonium-239 (along with other plutonium isotopes) is produced in nuclear reactors containing uranium-238. Uranium fuel containing uranium-238 irradiated in a reactor for very short times creates only small amounts of plutonium, but the plutonium-239 abundance is high compared with the other isotopes of plutonium. Such plutonium, containing at least 90 percent plutonium-239, is called weapons-grade plutonium. Longer irradiation, such as in a nuclear power reactor, generates more plutonium, but also causes undesirable plutonium isotopes to build up, so that the plutonium generates more heat, neutrons, and other radiation, complicating weapon design. Declassified documents have, however, disclosed that even reactor-grade plutonium can be used to build a nuclear weapon.[14]

Nuclear fuel reprocessing facilities can separate the portions of irradiated nuclear fuel that can be recycled (including plutonium) into new reactor fuel from waste products that are unwanted in new fuel. Reprocessing of spent nuclear fuel to separate and recycle the plutonium (and, in some schemes, other constituents) into reactors as fuel is sometimes referred to as closing the fuel cycle. A plant that reprocesses irradiated nuclear fuel or targets to separate plutonium may serve either a civilian nuclear energy program or a nuclear weapons program, or both. Some separations processes do not have material streams of direct-use nuclear material. Any separations process can be adapted to separate one or more of its constituents. The central questions in evaluating the proliferation aspects of a reprocessing facility are, (a) How much would that facility, when operational, reduce a country's time or cost for making a nuclear weapon illicitly, and (b) How easily could such illicit behavior be detected by the concerned outside world? These points are discussed in more detail later in this report. It is easier to detect reprocessing activities than enrichment activities, but clandestine programs for reprocessing irradiated nuclear fuels or special targets are also possible.

Today a relatively small number of countries enrich uranium (Brazil, China, France, Germany, India, Japan, Netherlands, Pakistan, Russia, the United Kingdom, and the United States have operating facilities, and Iran is trying to bring a new facility online). Some of these primarily serve weapons programs, and some others only serve national nuclear fuel needs.[15] Only two nations—Russia and the United States—and two international consortia—Eurodif (in France) and Urenco (in Germany, the Netherlands, the United Kingdom, and soon the United

---

[13] If the enrichment tailings (the depleted uranium by-product of enrichment) contain 0.3 percent U-235, then it takes 114 separative work units (SWU) to enrich about 220 kg of natural uranium to produce 33 kg uranium at 3 percent enrichment. Enriching that same 33 kg from 3 percent up to 90 percent produces 1 kg of HEU using 79 more SWU. Producing 19 kg of 5 percent enriched LEU from natural uranium requires 137 SWU, and enriching that same 19 kg to produce 1 kg of HEU at 90 percent enrichment requires only 55 SWU. But the SWU requirements for producing HEU drop significantly further if the enricher is willing to waste U-235 by leaving more of it in the tailings.

[14] A nuclear bomb can be made with reactor plutonium (NAS, 1994; DOE, 1997).

[15] For a discussion of Brazil's enrichment program, see Cabrera-Palmer, B. and G. Rothwell, 2008.

States and a joint venture in France)—provide commercial uranium enrichment services for other countries.

A smaller set of countries reprocesses irradiated nuclear fuel: China, France, India, Japan, Pakistan, Russia, and the United Kingdom.[16] Only France, Russia, and the United Kingdom offer commercial reprocessing services to other countries, and only Russia provides options in which the radioactive wastes generated in spent fuel reprocessing may not be returned if that is stipulated in international agreements.

FUEL FABRICATION

It is important to note that reactor operators use manufactured fuel assemblies in reactors, not raw enriched uranium. Fuel manufacturing (or fuel fabrication) is a process separate from enrichment. Fuel fabrication facilities must create the equipment for making fuel needed for each specific reactor design, and even within a reactor design there are differences among individual reactors. Reactors that operate at high efficiency require fuel with variations specified for each fuel assembly. This manufacture is highly specialized, but clients of a uranium enrichment center, or the fuel center itself, could contract out for these services.

Fuel fabrication services, like enrichment services, are in a competitive market, but with important differences. While low-enriched uranium is an interchangeable commodity (product from different enrichers can be essentially interchangeable), fuel fabrication is specific to the reactor that will use the fuel, and the fuel design is the intellectual property of the designer. Essentially all fuel fabricators compete to produce and sell fuel reloads for all reactors, whether the reactor is of their company's design or not, or even of the same reactor type (companies that sell pressurized water reactors also compete to sell fuel for boiling water reactors). Most fuel fabrication facilities today are located in the countries that have reactor vendors, which also mostly are the countries that enrich uranium. The fuel fabrication companies may be private corporations or state-owned (or quasi state-owned) corporations operating in the market, but all are subject to the laws, regulations, and policies of their governments. As with other parts of the nuclear fuel cycle, governments are able to block supply of fuel fabrication services. If a fuel fabrication facility is located in the United States, the U.S. government can approve or prevent that company's provision of nuclear fuel to a company in another country.

Manufacture of LEU fuel is not a proliferation problem, so a country worried about fuel manufacture for its own reactors as a link in the chain of assurance of supply could develop its own fuel fabrication facility (as South Korea has). For an international fuel center, whether fuel fabrication is offered is a question of economics, the marketplace, and the attractiveness of bundled services. For reactor types that have more than one fuel supplier, if assurance is desired without a domestic fuel fabrication capability, then agreements could be put in place where one fuel fabricator supplier can back up the other (IAEA, 2007d). If there are reactors with only one supplier, that has to be dealt with separately.

In summary, the spread of uranium enrichment technologies is a concern because of the following: (a) It is possible to design very simple bombs based on HEU (see, e.g., NRC, 2002). (b) Clandestine production of HEU using covert enrichment facilities is harder to detect than clandestine production of plutonium from covert reprocessing spent nuclear fuel. (c) Uranium enrichment services are needed to make nuclear fuel for almost any reactor, so a nation can more

[16] The Democratic People's Republic of Korea has a reprocessing plant that is now disabled.

readily argue that a uranium enrichment facility is part of its civilian nuclear energy program. Spent nuclear fuel reprocessing is a concern because a range of plutonium isotopic compositions (and even plutonium compositions containing minor actinides[17]) either is or can readily be made into direct-use material for a nuclear explosive device.[18] More compact, higher yield nuclear weapons can be made with plutonium. And experience working with plutonium (chemistry and metallurgy) is directly relevant to the manufacture of nuclear explosives. Thus, both technologies, uranium enrichment and spent fuel reprocessing, are of concern.

A broad range of institutional and technical measures to help stem the spread of nuclear weapons has been built up over several decades. The Treaty on the Non-Proliferation of Nuclear Weapons (NPT) is the foundation of this global regime, with IAEA safeguards providing the treaty's verification. All NPT parties that do not have nuclear weapons commit never to acquire them and to accept IAEA safeguards on all their civilian nuclear activities, while the NPT nuclear weapon states (China, France, Russia, the United Kingdom, and the United States) commit to negotiate in good faith toward nuclear disarmament, and all parties maintain their "inalienable right" to the peaceful use of nuclear energy. The NPT is complemented by national systems of exports control in the countries which possess relevant nuclear weapon technologies.

The financial costs for a nation to develop its own uranium enrichment and/or spent nuclear fuel reprocessing technologies or acquire them from elsewhere may be an economic constraint for a country pursuing the development of a national nuclear power industry. Moreover, there are technical barriers, including the acquisition of both required special knowledge and experience, and respective equipment and material relevant to the creation of nuclear explosive devices.

The U.S. and Russian governments have both supported the growth and spread of nuclear energy, but also have undertaken extensive efforts to limit the proliferation risks that might result if that growth and spread were not appropriately managed. In his speech to the United Nations Millennium Summit in 2000, then-Russian President Vladimir V. Putin called for steps to "reliably block the ways for spreading of nuclear weapons," including new approaches to the nuclear fuel cycle that would exclude "usage of enriched uranium and pure plutonium in world atomic energy production" (Putin, 2000). More recently, as discussed in this report, Russia has proposed the creation of a series of international centers for implementing key aspects of the nuclear fuel cycle, giving all countries the right to acquire the services of these centers and to participate in managing them and profiting from them, without spreading technology that could contribute to nuclear weapons programs. Russia has established the International Enrichment Center at Angarsk as one element of the network of centers it envisions. Russia has also focused on developing technologies for spent fuel regeneration and recycling in fast reactors that Russia believes pose reduced proliferation risks, as well as factory-built, sealed reactors, with long-life cores that could be provided to a country and then be removed, with their fuel, when their useful lives were done, so that the host country would not need to establish an extensive nuclear infrastructure of its own. (These are discussed in the second half of this report.)

The United States, similarly, has undertaken a broad range of efforts, in some cases working with Russia and with other governments, to limit the proliferation risks of nuclear

---

[17] The actinides are a group of heavy elements that includes thorium, uranium, and plutonium, as well as most of the products of neutron activation of those elements. Those products, such as americium and cesium, are called minor actinides.

[18] In August 1977, a memorandum from the Oak Ridge Laboratory described a "simple and quick" reprocessing facility. For an analysis of that suggestion, see GAO, 1978.

energy. Many of these steps were outlined in President George W. Bush's 2004 address on limiting nuclear proliferation risks (Bush, 2004). Steps the United States is currently pursuing include, among others: (a) combining new assurances of fuel cycle supply with other nuclear energy assistance into an "attractive offer" to be made jointly by the major suppliers to give countries establishing nuclear power programs incentives not to risk investing in their own enrichment and reprocessing capabilities; (b) working cooperatively with many of the states that are considering establishing nuclear power programs to ensure that they establish appropriate safety, security, and safeguards infrastructures and nonproliferation policies before their first reactors are built; (c) developing the international portion of the Global Nuclear Energy Partnership (GNEP), which is intended to bring major supplier and recipient states together in a common approach to nuclear energy in which many states could enjoy the benefits of nuclear energy without needing their own enrichment or spent fuel management facilities; (d) greatly expanding efforts to strengthen the IAEA and its safeguards system, in part under the rubric of the Next-Generation Safeguards Initiative, including pushing for an increased budget for the IAEA, additional provision of information and analytical support to the IAEA, research and development on more advanced safeguards technologies, and recruitment and training of appropriate experts to replenish the pool of safeguards experts; and (e) pursuing research and development on next-generation reactor and fuel cycle systems designed to have increased proliferation-resistance. How successful these efforts will be in reducing the potential proliferation hazards from the spread of nuclear energy remains to be seen.

Assurance of nuclear fuel supply for countries pursuing development of their own nuclear power industry is meant to minimize the incentive to develop their own technologies of uranium enrichment. Furthermore, in case of nuclear fuel supply in the form of nuclear fuel assemblies for such countries' nuclear reactors with a condition of spent nuclear fuel take-back (nuclear fuel leasing), the risk that these countries will acquire plutonium from spent fuel reprocessing is also lower. However, as the weapon states continue to enrich uranium and reprocess separated plutonium, some nonnuclear-weapons states (NNWS), both those that possess uranium enrichment and plutonium separations technologies and those that do not, have challenged any restriction on their right to do the same as creating another unfair divide between nuclear "haves" and "have-nots." But new structures of incentives have the potential to reduce the number of states that choose to invest in such facilities in the future (see, e.g., ISAB, 2008).[19]

Current proposals for fuel assurances and international incentives both seek to reduce the perceived risk that countries might cut off other countries' fuel supply for political reasons. The joint supply assurances from the major suppliers would help reduce this risk because countries could be confident they would have supply unless all of the major suppliers jointly decided not to supply them. The proposed fuel banks, if designed appropriately, would reduce this risk further, providing supply even if none could be had from any of the major suppliers. (To achieve that objective, however, it will be important to design such reserves so that the existing suppliers are not seen by recipient states as being able to readily prevent the reserves from providing supply.) The Angarsk enrichment center, Russia argues, would reduce the risk of a political interruption compared to states simply contracting for supply from Russian enrichment enterprises, because members of the center would have a government-to-government agreement prohibiting Russia from interrupting supply for political reasons. (How much confidence states would have in these

[19] The current major supplier states have talked about an "attractive offer," in which the IAEA, with support from those states, would assist countries in acquiring reactors, nuclear fuel supplies, and services. This assistance could be both technical and financial.

agreements remains to be seen.) Ultimately, however, in a world of sovereign states, the possibility that suppliers would decide not to supply would always remain, just as it does for other economically critical commodities and products, from oil to integrated circuits.

An approach that has not been utilized beyond the Soviet Union and now Russia's fuel contracts, but could have nonproliferation advantages, would be to offer full fuel services: uranium, enrichment, fuel fabrication, and perhaps take-back. This arrangement is described by Steve Kidd of the World Nuclear Association (Kidd, 2007) and is being explored by Russia.

Meanwhile, innovative reactor technologies applying a closed fuel cycle are being investigated. Rather than separating nearly pure plutonium and uranium from spent fuel, these new processing technologies generally keep some portion of the minor actinides and, in some cases, a portion of the fission products with the plutonium as it is recycled, with the goal of making the material in the fuel cycle more radioactive and less attractive for use in weapons (though in some proposals the difference might be small enough to have only a modest nonproliferation or counterterrorism benefit). The successful development and deployment of such technologies for peaceful nuclear power programs may reduce the risk of nuclear weapons proliferation compared with deployment of existing fuel cycles that recycle separated plutonium in mixed-oxide (MOX) fuel. The U.S. committee members believe that some of the processes under investigation might still produce material streams that do not require additional complex, remotely operated separations to extract direct-use material. Other processes might make extraction of direct-use material significantly more challenging. Even then, however, a set of safeguards would be required that can detect diversion in the operation of power-generating units of a nuclear power station, which are based on these innovative technologies. Further, clear goals for this institutional and technical system will be required to evaluate how much a particular technological system is contributing to nonproliferation goals.

As new countries engage in nuclear power activities, it will be important to strengthen IAEA oversight by convincing these countries to adopt the additional protocol and giving the IAEA additional resources, information, and authorities (IAEA, 2008), and to develop strong export and import control regulations to block development of a black market in nuclear materials.

While signing on to the NPT entails a commitment by nonnuclear weapons states to not develop nuclear weapons, the treaty contains no explicit prohibition of developing enrichment and reprocessing capabilities, and the IAEA has stated that such facilities are within the activities permitted to a nonnuclear weapon state under the NPT. A broad array of nonnuclear weapon states, including Argentina, Brazil, and South Africa, which are prominent voices within the world nuclear energy sector, ardently support the IAEA position.

As noted above, the director general of the IAEA, former Russian President Vladimir V. Putin, and U.S. President George W. Bush have proposed institutional arrangements that could assure the supply of needed nuclear fuel for countries having or planning to build nuclear power plants. These arrangements are intended to remove an incentive for countries to construct their own enrichment facilities, and, as noted above, to pursue several other nonproliferation goals. This is not a new concept. The idea of international ownership and management of sensitive nuclear-power-related facilities was first considered at the dawn of the nuclear age, going back to the immediate post-World War-II era and the Baruch Plan of that time (Baruch, 1946), and various options for international fuel banks or international fuel cycle centers have been discussed for decades. A few multinational fuel cycle enterprises have in fact been established, as discussed in the remainder of this report, demonstrating that multilateral or international fuel cycle centers are feasible.

The joint committees' statement of task explicitly focuses on international fuel supply centers, former Russian President Vladimir V. Putin's proposed approach. Questions concerning fuel cycle centers are discussed in detail below, but to identify and analyze the strengths and weaknesses of this approach, the joint committees also describe and examine the existing system of fuel supply and other proposed approaches that might be deployed in conjunction with fuel centers or serve as alternatives to them.

Finding 1b

Uranium enrichment and spent fuel reprocessing are the key technologies that enable countries to produce direct-use materials for nuclear weapons. The more countries to which either technology (enrichment or reprocessing) spreads, the greater the proliferation risks. Currently it appears that more countries that have not already deployed these technologies are interested in establishing uranium enrichment programs than in pursuing spent fuel reprocessing technologies, making the spread of enrichment technology a greater near-term concern for nuclear proliferation. But the intention to acquire spent nuclear fuel reprocessing capabilities was the main focus of proliferation concerns in the 1970s and could become so again.

Finding 1c

Requirements of the nuclear security environment, the difficulty of providing safeguards and security, and the demand for nuclear fuel cycle services change over time, and technology advances with time. Any approach for enhancing the nonproliferation features of international fuel cycles must be staged to respond to the nonproliferation needs of the time period. Today this suggests a focus on convincing countries that they do not need to establish their own enrichment facilities, which has motivated efforts by several countries and international organizations to address the enrichment issue. Similar efforts are needed to convince countries that they do not need their own reprocessing facilities. Also needed are strengthened efforts to prevent the spread of these technologies through illicit or inadequately regulated exports and black-market nuclear networks, and improved safeguards for both uranium enrichment and spent fuel reprocessing facilities, designed both to increase international confidence that significant diversions from declared facilities would be detected and to strengthen the ability to provide timely warning concerning covert facilities and activities.

Recommendation 1a

The countries that currently provide nuclear fuel services should redouble efforts, with other countries and the IAEA, to establish mechanisms for increasing reliability of supply of nuclear fuel, so that countries that do not now have enrichment technology would have reduced incentives to build their own uranium enrichment facilities.

Recommendation 1b

The international community should help countries provide adequate capacity for safely storing spent fuel (on their own territory or elsewhere), or reliable reprocessing services from existing providers, to reduce countries' incentives to establish their own reprocessing

facilities. Separated plutonium or fabricated plutonium fuel should not be sent to countries that have not previously received such material and do not have reprocessing capabilities. The spread of separated plutonium to additional countries poses many of the same proliferation risks posed by the spread of reprocessing capabilities.

Recommendation 1c
For similar reasons the United States and other nations should reduce and seek to minimize commerce in and the transfer of highly enriched uranium (which poses proliferation risks) except if sealed in a reactor core.

Second-level findings:

a. **To ensure a reliable supply of nuclear fuel, a country needs reliable fuel fabrication services as much as it needs reliable sources of uranium and enrichment services.**
b. **To assist in the international fuel assurance programs, it would be helpful if nations with fuel fabrication facilities made those available.**
c. **Fuel fabrication technology for uranium oxide fuel with low-enriched uranium is not sensitive from a proliferation perspective. Hence, if countries choose to establish their own fabrication capabilities to produce fuel assemblies for their own nuclear power stations, without establishing uranium enrichment or spent fuel reprocessing capabilities—as South Korea has done, for example—this should not pose significant international concerns.**

Finding 2
Several messages are clear from the NAS-RAS Workshop and other recent discussions in Vienna about assurance of supply:

a. **Few countries have declared a willingness to forgo forever a right to develop their own uranium enrichment or spent fuel reprocessing nuclear technology in the future. Some countries have expressed adamant opposition to requiring a country to forgo the development of its own enrichment and reprocessing technologies as a condition of assurance of supply of nuclear fuel or low-enriched uranium.**
b. **To be successful, uranium enrichment, fuel assembly production for nuclear power stations, and spent fuel storage and reprocessing continue to operate in the international market.**
c. **No single mechanism or strategy for assurance of nuclear fuel supply is likely to address every country's legitimate needs and desires. Each country's or region's needs and requirements may be different.**
d. **New mechanisms for assured nuclear fuel supply may only modestly change countries' incentives to establish enrichment facilities, as the existing international market provides strong assurance of supply and countries have a variety of other reasons for establishing their own enrichment plants, including a desire to participate in the profits of enrichment, national pride, and a desire to establish a nuclear weapons option for the future.**

Introducing management-based or technology-based mechanisms to inhibit or limit the spread of enrichment (and reprocessing) facilities must be handled carefully to avoid increasing the likelihood of new states establishing domestic enrichment. Discussion of restricting access to enrichment technology, even with international fuel supply centers, has prompted more countries (not fewer) to declare their interests in developing enrichment facilities within their borders.

Recommendation 2a

The governments of the United States and Russia should continue to support a broad menu of approaches to increasing assurance of nuclear fuel supply.

> **An array of mechanisms for assurance of nuclear fuel supply has been proposed, from diversified long-term contracts through the existing market, enrichment bonds,[20] and international fuel centers to creating a virtual or actual fuel bank. Some of these are already in place. The Russian and U.S. governments should support a broad menu of these approaches, ensuring that these do not undermine each other.**

Recommendation 2b

The governments of the United States and Russia should seek to establish additional benefits and incentives for countries that choose not to establish their own uranium enrichment and spent fuel reprocessing facilities. Possibilities could include assistance with establishing the necessary infrastructure for safe and secure use of nuclear energy.

Recommendation 2c

To support nonproliferation goals, the nations that currently supply nuclear fuel should work expeditiously with other countries and the IAEA to make assured fuel supplies available before there is a major commitment to new nuclear power plants by countries that do not have them today.

[20] Enrichment bonds: A guarantee by a state that supplies enrichment services that enrichment providers will not be prevented from supplying the recipient state with uranium enrichment services if the guarantee is invoked (adapted from the U.K. proposal).

CHAPTER 2

INTERNATIONAL NUCLEAR FUEL CYCLE CENTERS

A.1 Is it feasible and effective to establish international nuclear fuel supply centers as an incentive for countries not to develop indigenous enrichment facilities?

The joint committees were asked, Is it feasible and effective to establish international nuclear fuel supply centers as an incentive for countries not to develop indigenous enrichment facilities?

It is, indeed, feasible to establish international enrichment centers, as demonstrated by the fact that two such centers exist and Russia is creating another one. Urenco represents one approach, where each of the partners (Germany, the Netherlands, and the United Kingdom) has an enrichment facility within its borders, and shares knowledge of the centrifuge technology. New partners to Urenco, France and the United States, will not have access to the technology. Eurodif operates a facility in France, and its partners (Belgium, Spain, and until 1974 Sweden)[1] obtain enrichment services from the Eurodif facility; while the partners serve on the decision-making board, they do not help operate the facility and have no access to the technology. Russia is establishing a center at Angarsk with joint ownership by other countries, similar in some respects to the Eurodif approach; in particular, foreign partners will not participate in facility operations and will have no access to the technology. Russia has said that joint facilities for other fuel services could be set up on its territory in the future.

In 2006, the French nuclear group, AREVA, entered into a joint venture with Urenco, the joint British-Dutch-German uranium enrichment centrifuge consortium, acquiring a 50 percent share of ETC, the Enrichment Technology Company, which comprises all of Urenco's centrifuge design, manufacturing, and related research and development. Despite owning a 50 percent share of ETC, France does not have a right-to-access to ETC's centrifuge technology. ETC is providing centrifuges to AREVA's new enrichment facility, Georges Besse 2, located in Tricastin, France, and to the National Enrichment Facility (NEF) located in New Mexico in the United States, led by Urenco. In both cases, the centrifuges will be in "black boxes" so that neither French nor U.S. personnel will have access to the centrifuge technology—though factories may be built in France and the United States, staffed by ETC personnel, to produce centrifuges. The ETC required intergovernmental agreement between the governments of France and Germany, the Netherlands, and the United Kingdom to set up a joint venture.

Eurodif is a joint stock company formed by Belgium, France, Spain, and Sweden in 1973. Sweden withdrew from the company in 1974 and was replaced by Iran and later by Sofidif, a joint French-Iranian venture. Eurodif's operating facility, Georges Besse 1, uses gaseous

[1] Iran is also a partner, through Sofidif, as described later.

diffusion technology to enrich uranium for nuclear power utilities that operate nuclear power plants, including EDF. The Georges Besse 2 plant will replace Georges Besse 1.

At a meeting of the Interstate Council of the Eurasian Economic Community on January 25, 2006, the Russian President Vladimir V. Putin proposed the creation of a network of international nuclear fuel cycle centers to provide "nuclear fuel cycle services, including enrichment, on a non-discriminatory basis and under control of the [International Atomic Energy Agency] IAEA," (IAEA, 2006).[2] To implement this proposal, the International Uranium Enrichment Center (IUEC) was set up on the site of the Angarsk Electrolysis Chemical Complex (AECC) with the aim of providing "IUEC-participating organizations with guaranteed access to uranium enrichment capabilities," (IAEA, 2007a). The main principles underlying IUEC development are as follows (Ruchkin and Loginov, 2006):

- The center will be a commercial organization and operate as an open, joint-stock company supervised by a joint advisory committee (with IAEA representation).
- All countries not pursuing the development of weapon-related sensitive nuclear technologies and meeting all nonproliferation requirements will be eligible for equal, nondiscriminatory IUEC membership.
- Russia maintains national control over the material, and export regulations will be developed to guarantee shipment of the material to any participating state at their request, or to other states at the IAEA's request.
- Part of the AECC's production facilities will be made eligible for voluntary IAEA safeguards.[3]
- Participants will have no access to Russian uranium enrichment technology.
- Enriched uranium should meet the requirements of nuclear power stations for nuclear fuel for participant countries.
- The political, economic, and technological advantages to IUEC membership should outweigh the drawbacks of refraining from full nuclear fuel cycle development.

On May 10, 2007, the head of the Russian Federal Atomic Energy Agency, Sergey Kirienko, announced that five to seven countries had expressed interest in joining the IUEC. Through the signature of an intergovernmental agreement on that day, Kazakhstan then became the first joint member. Armenia and the Ukraine have expressed interest in joining.[4] In the future, international centers could be developed and set up for spent nuclear fuel management (including its long-term storage and reprocessing and further use in innovative fast reactors), innovative reactor and nuclear fuel cycle technology development, or nuclear personnel training (Ruchkin and Loginov, 2006).

Russia is discussing with IAEA a mechanism enabling shipment of material out of Russia at IAEA request, which might contribute to a broader IAEA structured assurance of supply.

---

[2] In Russian, the word *kontrol'*, rendered here as "control," often refers to monitoring rather than actual management of a facility. Russia has not made any proposal that the IAEA should manage the Angarsk enrichment enterprise.

[3] Which part of the facility eligible for safeguards is still being worked out between the AECC and IAEA.

[4] For a chronology of events regarding the Angarsk International Uranium Enrichment Center Chronology, see PIR Center for Policy Studies (Russia), http://pircenter.org/index.php?id=1976&gfkey=chronology.

WHAT IMPACT CAN SUCH CENTERS AND ASSURED FUEL SUPPLY IN GENERAL HAVE ON NONPROLIFERATION ISSUES?

Whether a facility is under national, multinational, or international control need not have a major effect on its role in the international commercial marketplace. Urenco, for example, is a multinationally controlled enrichment enterprise that provides enrichment services both to its partner countries and to other countries on a commercial basis. Future multinational or international centers might do the same. Indeed, if an existing nationally controlled facility were converted to multinational or international control, its role in providing enrichment services internationally might be much the same as it was before. If, in the future, governments decided to subsidize the establishment of an internationally controlled enrichment facility, perhaps for nonproliferation reasons, decisions would be needed as to how this facility would relate to other players in the commercial marketplace.

Fuel supply centers are one of several possible options for assurance of supply of nuclear fuel. IAEA Director General Mohamed ElBaradei and a working group of the IAEA Secretariat submitted to the IAEA Board of Governors in June 2007, a report titled *Possible New Framework for the Utilization of Nuclear Energy: Options for the Assurance of Supply of Nuclear Fuel* (IAEA, 2007d). The report lays out a multilayered and multilateral approach to assuring supply of nuclear fuel against political disruptions.[5] "The risk of such disruptions might dissuade countries from initiating or expanding nuclear power programmes and/or create vulnerabilities in the security of supply of nuclear fuel that might drive States to build their own national enrichment capabilities with possible additional proliferation risks" (IAEA, 2007d). Mechanisms for assurance of fuel supply, then, provide countries an incentive for developing nuclear power without developing their own capacity for uranium enrichment, particularly if countries can only exercise the mechanism if they are not conducting enrichment activities.

Incentives, by definition, reduce rather than eliminate the risk of a determined nation developing domestic enrichment facilities for reasons of national pride or seeking nuclear weapons capabilities. The incentives can, however, increase the degree of resolve required for a country to take this step by making it less attractive from economic and political perspectives (see the next section concerning economic aspects of these questions). Providing credible offers of assured fuel supply on attractive terms could also help focus international attention on the motives of countries that rejected these offers. While it cannot be assumed that a nation rejecting such offers aspires to nuclear weapons capability, the availability of a mechanism for assurance of fuel supply undercuts that particular argument and strengthens suspicions that the country may be trying to develop the option of a nuclear weapons program. Mechanisms other than assurance of fuel supply, such as nuclear fuel leasing with spent fuel take-back, may be possible and could prove to be significantly stronger incentives against developing enrichment capabilities than assurance of fuel supply. The United States and Russia are also working on forms of assistance such as infrastructure planning and development, financing, and linkage of reactor supply as deterrents to developing enrichment for now. As a current example, although not a leasing agreement, Russian supply of nuclear fuel to a nuclear power station in Iran is carried out on the

[5] The report defines political disruptions as disruptions unrelated to technical or commercial considerations. Because the assurance of supply is meant to serve nonproliferation objectives, it could not be exercised to compensate for disruptions related to safeguards violations or nonproliferation transgressions. This IAEA report built on the considerations of an earlier expert group established by Director General ElBaradei and chaired by former Deputy Director General Bruno Pellaud (IAEA, 2005b).

condition of spent fuel take-back to Russia. Assurance of supply and fuel leasing are both discussed in more detail below.

Assurance of fuel supply can itself mean several different things. At our workshop, some participants argued that such a mechanism should mitigate not only political disruptions of supply, but any disruptions of supply. This argument has not, to date, gained much support, because the existing fuel supply market works well and nonpolitical disruptions are viewed as fairly unlikely. Reactor operators already use a variety of mechanisms to reduce risk of interruption of supply, such as backup contracts with different suppliers and stocking fuel reserves to assure themselves that fuel will be available. Indeed, the IAEA working group on assurance of fuel supply established early on that any proposed mechanism for assurance of supply should not disrupt the existing market, for fear of damaging a system that is functional and reliable.

The IAEA proposal is for a multilayered assurance of supply that would include primary reliance on the commercial market, commitments from suppliers to provide backup supplies if politically motivated interruptions occur, and one or more fuel banks as a final layer of assurance. This proposal was based in part on a proposal for "reliable access to nuclear fuel supply" (RANF) developed by the major suppliers.[6]

The 2007 IAEA document mentioned above (IAEA, 2007d) describes a structure for assurance of nuclear fuel supply that would operate as a tiered set of mechanisms, with the existing market as the first tier, a virtual fuel bank or enrichment bonds as a second tier,[7] and an actual fuel bank as the third tier, to be exercised only if the first two fail. Countries, under this proposal, would have access to these mechanisms based on four possible criteria for states to be able to access the proposed assured fuel supplies: (1) that the disruption is political; (2) that the state have a safeguards agreement in force for the material;[8] (3) that the state be in good standing with respect to its safeguards commitments, with no issues before the IAEA Board of Governors; and (4) that the state comply with other criteria that may be imposed by the Board of Governors (such as having an additional protocol in force). One additional criterion discussed in the proposals from the major suppliers has been that the nation receiving supply assistance may not currently be engaged in enrichment activities.

One could envision a slightly different approach, again using a set of tiered mechanisms, that offers different types of assurance based on the different levels of nonproliferation and sensitive-technology commitments made by the participating nations. Specifically, if the international community were to offer the additional benefit or incentive of assurance against nonpolitical disruptions, a criterion, condition, or payment for that service might be based on the level of a nation's commitment not to enrich uranium or reprocess nuclear fuel. For example, any nation that signs an agreement to not develop enrichment (not forever, but perhaps 10 years or 20 years) could access a fuel bank in cases of *any* interruptions of supply (or possibly even exorbitant price spikes) that are not the result of the country violating nonproliferation agreements. Countries could offer this additional commitment regarding sensitive technologies as a form of payment for the assurance of a supply mechanism that would provide against normal

[6] The six enrichment-services-supplier states set up an intergovernmental working group to develop a Concept for a Multilateral Mechanism for Reliable Access to Nuclear Fuel (RANF), signed by France, Germany, the Netherlands, Russia, the United Kingdom, and the United States in 2006.

[7] A virtual fuel bank is often defined as a commitment by one or more fuel suppliers to provide fuel if called upon by IAEA, but is not a dedicated separate stock of fuel.

[8] Note: This does not mean that full-scope safeguards would be required.

market disruptions. As noted above, reactor operators now pay for other insurance mechanisms (backup contracts and backup onsite inventories of fresh fuel),[9] so a monetary value can be placed on the service such an assurance entails.

Some nations, notably some in the Non-Aligned Movement,[10] are cautious about mechanisms for assurance of supply and critical of additional criteria for accessing them. From this perspective, whatever advantage is offered by a fuel bank, for example, is reduced if the suppliers can deny access. So if the Russian government's approval is needed to release the enriched uranium Russia pledged to the IAEA fuel bank (Kirienko, 2007), and if the U.S. government must agree in each case to release the enriched uranium it has pledged, these major suppliers still control whether fuel is supplied through the fuel bank. Being among the owners of an enrichment facility, and having a government-to-government agreement in place prohibiting any interruption of supply from that plant, may significantly increase states' sense of assurance about fuel supply, although being part owner of Eurodif has not allowed Iran to access Eurodif services.

The assurance of nuclear fuel supply could mean assurance of access to uranium enrichment services or to uranium enrichment and fuel fabrication services, or it could mean access to a stock of material, uranium as uranium hexafluoride (UF_6) or uranium oxide powder (U_3O_8). Creating a stock of fabricated fuel is less feasible because of the reactor-specific features and characteristics of fuel and fuel elements. Nuclear fuel is highly specialized, and each nuclear power station reactor needs nuclear fuel with inherent specific characteristics of this reactor.

International fuel supply centers are somewhat different from such mechanisms as virtual or real fuel banks. Fuel supply centers could be structured to operate entirely within the existing enrichment market, providing only its joint stockholders with assurance against disruptions of supply. The incentives for not developing an enrichment or reprocessing facility would extend only to those joint stockholders. Assurances might also be extended to contracting parties, which would broaden the potential effect of the center. In both cases, substantial work would be needed on legal questions to establish enforcement mechanisms for exercising the assurance mechanism against the political will of the (other) joint stockholders if they are the cause of a political disruption of supply.

A very different approach would entail long-term contracts between each joint stockholder and the center, which could insulate the participants against price fluctuations but would make the participants reliant on the center's performance. To serve as an incentive, the center would need to be economically competitive, factoring in whatever benefits a nation perceives from the assurance of fuel supply.

---

9 Reactor operators do not typically purchase a substantial reserve stock of fuel because of the cost of having that capital sitting idle and unproductive, but some operators do choose this option. South Korea has enough fuel and enriched uranium to supply its reactors for 1 year, but this is partly because South Korea fabricates its own fuel and has material and fuel in its pipeline.

10 "The Non-Aligned Movement is a Movement of 115 members representing the interests and priorities of developing countries….[The Movement attempts] to create an independent path in world politics that would not result in Member States becoming pawns in the struggles between the major powers. [T]hree basic elements which influenced the approaches of the Movement to international issues…are the right of independent judgement, the struggle against imperialism and neo-colonialism, and the use of moderation in relations with all big powers." *The Non-Aligned Movement: Description and History* (http://www.nam.gov.za/background/history.htm).

ECONOMIC ASPECTS OF THE NUCLEAR FUEL CYCLE

Fresh LEU Supply

Fresh fuel typically contributes less than 10 percent of the cost of nuclear-generated electricity today.[11] Each component of low-enriched uranium (LEU) fuel supply—the production of natural uranium, conversion to uranium hexafluoride, enrichment, fabrication into fuel assemblies, and delivery of the fuel to the reactor—is characterized by a mature and competitive market. Thus, it should be possible to construct fuel supply assurances that have very little effect on the cost of nuclear-generated electricity and the functioning of markets for fuel services, and which do not disadvantage either suppliers or recipients. Consider, for example, an international store of low-enriched uranium that is created through donations from supplier countries, such as the United States and Russia. A country or reactor operator that draws from this store under agreed rules could be required to pay the prevailing market prices for the natural uranium and separative work used to produce the withdrawn material. The method for setting the market price could be fair to both suppliers and their customers. One possibility would be to set prices equal to those that would be paid to replenish the stock, which would have the virtue of making the international fuel bank automatically self-sustaining. (The fuel bank might seek to have in place at all times a contract to supply enriched uranium.) Alternatively, the prices could be set equal to contract prices that had already been negotiated between the customer and the original supplier. In either case, the impact on the cost of electricity or the operation of markets for fuel cycle services would be small.

Spent LEU Take-back

The take-back of spent fuel is substantially more complicated because there are several possible options for the disposition of spent fuel. Moreover, because there is a lack of competitive markets for most services at the back end of the fuel cycle, the costs of the various options are uncertain. These services include long-term spent fuel storage and disposal, and reprocessing of spent fuel followed by disposal of the resulting wastes and fabrication of fuels for recycling or transmutation.

Spent Fuel Transport and Storage. Of the back-end services, the costs of long-distance transport and long-term storage of spent fuel are relatively low and well established.[12] The cost of spent fuel transport is on the order of $70-100/kg for transcontinental shipment by truck or rail (Shropshire et al., 2008); the cost of intercontinental transport by ship may be as high as

---

[11] Although spot prices for uranium have been volatile recently (rising substantially and dropping somewhat), nuclear fuel is usually procured through long-term contracts. Fuel cycle costs, including waste disposal, historically have been taken to be 10 percent of the cost of electricity, (see, e.g., IEA 2007). A more recent study by the U.S. Congressional Budget Office states that "Doubling [the fuel cost of nuclear power] would increase the levelized cost of new nuclear capacity by about 15 percent above that assumed in the reference scenario." (CBO, 2008, Chapter 3.6.1) This suggests that the fuel costs may be 15 percent of the estimated cost of electricity, which is $8/MWh (in 2006 dollars) based on long-term projections by the U.S. Energy Information Administration and including $1/MWh to cover the cost of disposal. This would imply a fresh fuel cost of about 13 percent of cost of electricity.

[12] Regulatory, standards-setting, and waste handling organizations state that the transportation of nuclear fuel has an excellent safety record. See U.S. NRC, 2005; and DOE, Undated. A U.S. National Research Council committee reasoned that the driver qualifications, standards, and scrutiny for such shipments probably contribute to a better safety record than in transportation of other goods (NRC, 2006).

$200/kg.[13] The life-cycle cost of providing 50 years of dry storage is estimated at $100-300/kg (Shropshire et al., 2008, pp. E2-16). For comparison, the cost of fresh light-water reactor (LWR) fuel is $1500-3000/kg.[14] Thus, the cost of spent fuel take-back for long-term storage (for example, in an international spent fuel storage center) is relatively small—on the order of 10 percent of the cost of fresh fuel or 1-2 percent of the cost of nuclear-generated electricity.

If the number of countries willing to take back spent fuel is limited, the price charged for take-back services could be substantially higher than the cost of providing the service. The price that reactor operators would be willing to pay for spent fuel take-back is unknown in the absence of a market for this service, but prices as high as $1,000-1,500/kg have been discussed.[15] The prospect of correspondingly large profits could provide the incentive necessary for a country to provide take-back services—and ultimately stimulate additional countries to provide this service.

Direct Disposal. One option for the ultimate disposition of spent fuel is disposal in a deep geological repository. Several countries, including Finland, Sweden, and the United States have advanced programs for the geological disposal of spent fuel, but no repository has yet accepted spent fuel (or other high-level waste) for permanent disposal. The take-back of spent fuel for geological disposal is a possibility, but to date no country has indicated a willingness to accept foreign power-reactor fuel for direct disposal.[16] Regional or international repositories have also been discussed, but no country has indicated a willingness to host such a site.[17]

The total undiscounted costs of geological disposal have been estimated at $400-900/kg (Shropshire et al., 2008, pp. F1-10). In the United States, the cost of building and operating the Yucca Mountain repository is to be financed through a $1/MWh charge on nuclear-generated electricity, which is less than 2 percent of the total cost of electricity.[18]

Thermal Recycle. A second option for the disposition of spent LEU fuel is to reprocess the fuel and recycle the recovered plutonium in mixed-oxide (MOX) fuel for another pass through light-water or other thermal reactors.[19] France and the United Kingdom have reprocessed spent fuel from other countries, including Belgium, Germany, Japan, and Switzerland. In doing so, France and the United Kingdom have provided only the reprocessing service; the resulting plutonium in storage and high-level waste remain the property of the owner of the spent fuel, and by contract are to be returned for recycling (as MOX fuel) and disposal, respectively. The prices charged for these services have been estimated at $2,000-2,500/kg in

---

[13] Atsuyuki Suzuki, personal communication based on information accessed at http://www.meti.go.jp/policy/electricpower_partialliberalization/contentscost-rire on September 2, 3008.

[14] A cost of $1,500/kg corresponds to average contract prices in 2006 ($60/kg for uranium and conversion, $120/SWU for enrichment, $220/kg for fabrication); $3,000/kg corresponds to spot prices in early 2008 ($200/kg for uranium and conversion, $150/SWU for enrichment), for pressurized water reactor fuel with a burn-up of 50 MW_td/kg.

[15] See Bunn et al., 2001, pp. 73-77, which describes a $1,000/kgHM estimate from Pangea for a disposal service in Australia; an estimate of $1,500/kgHM for a proposed storage and disposal service in Russia; and estimates of $300 to $600/kgHM for temporary storage, or $1,200 to $2,000/kgHM for reprocessing with no return of wastes or plutonium.

[16] Russia has passed laws to enable it to accept foreign spent nuclear fuel for reprocessing, including disposal of the waste, and offers fuel services whereby Russia retains ownership and takes back the fuel. These points are discussed later in this section.

[17] Examples are the inability to secure approval for storage or repository sites for a Pacific Basin spent fuel storage facility on Palmyra Island and the Pangea attempt to develop a repository in Australia.

[18] A $1/MWh charge is equivalent to $400/kg of spent fuel, assuming a burn-up of 50 MW_td/kg and an efficiency of 33 percent. With interest, these payments are estimated to be sufficient to pay the cost of the construction and operation of the Yucca Mountain repository.

[19] Using light water reactors for recycle was examined in Collins et al., 2007.

the initial contract period (Bunn et al., 2003), and \$900/kg subsequently.[20] Costs as low as \$500/kg have been estimated for a new, large reprocessing facility in the United States (Boston Consulting Group, 2006).[21]

Of the countries listed above, all but the United Kingdom have used some of the recovered plutonium in MOX fuel for thermal reactors. The separation of plutonium has outpaced its use in MOX fuel, however, leading to large stocks of plutonium. The prices charged for MOX fuel fabrication are not publicly available, but are estimated to be \$1,200-4,000/kg (Bunn et al., 2003, p. 216). At the low end of this range, the cost of MOX fuel (ignoring the cost of recovering the plutonium) is less than the cost of fresh LEU fuel (including the costs of natural uranium and enrichment).

Because spent LEU fuel contains about 1 percent plutonium and fresh MOX fuel contains about 6 percent plutonium, thermal recycling can supply one-sixth of the fuel for a fleet of LWRs. The cost of thermal recycling is dominated by the cost of reprocessing, less any cost savings from decreased waste storage or disposal costs and reductions in fresh fuel supply.[22] Assuming, for purposes of illustration, a reprocessing cost of \$1,000/kg, storage/disposal cost savings of \$200/kg, and a MOX fabrication cost that is equal to the total cost of fresh LEU fuel, the additional cost of thermal recycling is \$2/MWh. Although there is significant uncertainty in each of these parameters, a reasonable range for the net cost of thermal recycling is \$1-2/MWh, equal to 2-4 percent of the cost of electricity and about \$7-15 million/yr per GW_e of capacity.

As indicated above, France and the United Kingdom have provided only the reprocessing service, with the return of the separated plutonium (once fabricated into MOX fuel) and high-level wastes to the customer. If, in keeping with the goal of limiting the spread of sensitive fuel cycle steps, the return of plutonium or MOX fuel to countries without sensitive fuel cycle technologies is prohibited, it seems unlikely that those countries would pay the additional costs associated with the reprocessing of their spent fuel, unless the take-back country also assumed responsibility for final disposal of the high-level wastes. But if the take-back country assumes responsibility for both the plutonium and waste, the decision of whether to reprocess for thermal recycling would be entirely up to the take-back country. Although a user country might be willing to pay a price high enough to cover the costs of reprocessing and waste disposal, the take-back country could reduce its costs and increase its profits through long-term storage or direct disposal, or by deferring reprocessing until the recovered plutonium could be used immediately and cost-effectively in reactor fuel.

Plutonium storage. A variant of the previous option is reprocessing followed by long-term storage of the separated plutonium. The Soviet Union (and, now, Russia) has reprocessed

---

[20] Press release, "EDF and AREVA sign a contract for managing EDF used nuclear fuel," August 24, 2004. Cited in Shropshire et al., 2008, pp. F1-10.

[21] The \$500/kg cost supposedly includes all operating and capital costs, including interest on borrowed money. Some members of the joint committees are skeptical that unit costs this low can be achieved in practice.

[22] To a first approximation, the increase in the cost of LWR electricity (compared to direct disposal) due to thermal recycle, ΔCOE (\$/MWh), is given by:

$$\Delta COE \approx \frac{C_{rep} - \Delta C_{disp} + \frac{\Delta C_{fuel}}{6}}{24B\varepsilon}$$

where C_{rep} is the unit cost of reprocessing, ΔC_{disp} is the unit cost savings due to the storage and disposal of high-level waste instead of spent fuel, ΔC_{fuel} is the unit cost difference between MOX fuel fabrication and the total cost of fresh LEU fuel of equal burn-up (all in dollars/kg of heavy metal in the fuel), B is the burn-up of the fuel (MW_td/kg), d is days, and ε is the thermal efficiency of the reactor (MW_e/MW_t).

spent fuel from other countries, including Armenia, Bulgaria, and Finland. Under Soviet fuel supply agreements, the fuel remained the property of the Soviet Union, including the plutonium and wastes separated during reprocessing. Russia continues to supply reprocessing services for foreign customers. Russia has accumulated a large stock of separated plutonium from the reprocessing of domestic and foreign commercial spent fuel. Rather than recycle this plutonium in LWRs, Russia plans to use the plutonium for start-up fuel for future fast-breeder reactors.

Safe and secure storage of plutonium is expensive. The Mayak storage facility in Russia, which has a design capacity of 100 metric tons of plutonium, cost $421 million to build (completed in 2003) and has an estimated operating cost of $13 million/yr.[23] The cost to build a new storage facility in the United States for 45 metric tons of plutonium has been estimated at about $600 million, with an operating cost of $75 million/yr (Shropshire et al., 2008). Assuming facility lifetimes of 50 years, the corresponding undiscounted life-cycle costs would be $12,000 and $100,000/kg of plutonium capacity, respectively. Because spent fuel contains 1 percent plutonium, this is equivalent to $120-1,000/kg of spent fuel. This can be compared to the $100-300/kg given above for the long-term storage of spent fuel. Long-term spent fuel storage has the added advantage of deferring reprocessing for several decades, which is equivalent to a cost savings of at least $300-400/kg.[24] Plutonium that is separated from recently discharged spent fuel still contains nearly all of its plutonium-241. If that plutonium is then stored for some years, the plutonium-241 decays with a 14.4- year half-life to produce americium-241—a radionuclide that complicates fuel fabrication and handling. If instead that spent fuel is stored for the same number of years, and the plutonium is separated soon before it will be fabricated into fuel, the in-grown americium will be separated and the plutonium will be more pure at fabrication. Restated, as a result of the in-growth of americium-241, plutonium separated from recently discharged spent fuel and then stored is more difficult and costly to fabricate into fuel than plutonium separated after storage.

Transmutation. A final option for the take-back of spent LEU fuel is that envisioned by the U.S. Global Nuclear Energy Partnership (GNEP): reprocessing followed by immediate transmutation of the recovered plutonium and other transuranics (TRU) in a fast reactor. In addition to the costs of reprocessing the spent LEU fuel discussed above, there would be costs associated with the construction and operation of the fast reactor, including the reprocessing and fabrication of TRU fuels. Because fast reactors can transmute TRU isotopes that are responsible for most of the long-term heat load from spent LEU fuel, transmutation may offer significant reductions in geological disposal costs.[25]

The required fast reactor capacity depends on the *conversion ratio* of the fast reactor, which is the average number of TRU atoms produced per TRU atom consumed or fissioned in the fuel. A fast reactor fissions about 880 kg/yr of TRU per GW_e of installed capacity.[26] A

---

[23] The construction cost includes the facility and all the containers (see http://www.nti.org/e_research/cnwm/securing/mayak.asp, accessed on December 12, 2008), and the operating costs are taken from Shropshire et al., 2008, pp. E3-5.

[24] Assumes a reprocessing cost of $500/kg, a discount rate of 3 percent per year, and a delay in reprocessing of 30-50 years. For discount rates higher than 7 percent per year, the net present value of deferring reprocessing is essentially the cost of reprocessing, which, as noted above, could be $1,000/kg or higher.

[25] Recycling the plutonium from irradiated LWR fuel once through an LWR as MOX offers no heat-load advantage compared to simply using LEU fuel for both cycles. Fast reactors are more efficient in fissioning nonfissile transuranic isotopes.

[26] Assumes 0.93 MW_td of energy released per gram of TRU fissioned, a net thermal efficiency of 38 percent, and a capacity factor of 85 percent.

LWR fueled with LEU discharges about 250 kg/yr of TRU per GW_e of capacity.[27] Thus, if no new TRU were produced in the fast-reactor fuel (a conversion ratio of zero), 250/880 = 0.28 GW_e of fast-reactor capacity could consume the TRU from 1 GW_e of LWR capacity. But proven fast-reactor fuels are mostly uranium, leading to the production of additional plutonium and other TRU in the fast-reactor fuel. Although the fuel and core design can be modified to minimize the production of TRU, there are limits to what can be achieved while maintaining an acceptable degree of safety. A conversion ratio of about 0.7 is achievable using existing reactor and fuel designs, which would lead to a net consumption of only 260 kg/yr of TRU per GW_e of fast-reactor capacity, in which case the installed fast-reactor capacity would be about equal to the installed LWR capacity for the fast reactors to consume all of the TRU from the LWRs. A goal of the U.S. Advanced Fuel Cycle Initiative is to achieve a conversion ratio of 0.25, in which case fast reactors would comprise 27 percent of total nuclear capacity.[28]

A thorough assessment of the economics of transmutation would require the development of detailed models involving dozens of parameters, most of which have very large uncertainties. Two points can, however, be made at this time.

First, the effect of transmutation on the cost of nuclear-generated electricity will depend largely on the capital and operating cost of fast reactors relative to thermal reactors. Fuel-related costs are a relatively small contribution to the cost of electricity from either type of reactor, and differences in fuel-related costs will almost certainly be small compared to differences in capital and operating costs. Limited experience with fast reactors in France, Japan, and Russia suggests that fast reactors are likely to cost more to build and operate than thermal reactors. Some people believe, however, that a new generation of improved fast reactors could have significantly lower costs than next-generation light-water reactors, in which case transmutation would decrease the average cost of nuclear electricity. Because the technologies required for transmutation have not yet been specified, there is little factual basis today for judging whether separation and transmutation would ultimately increase or decrease overall costs.

Second, regardless of whether fast reactors are more or less expensive than thermal reactors, transmutation would require a mechanism to pay for the extra costs of the more expensive component of the system. If fast-burner reactors are more expensive than thermal reactors, a mechanism would be needed to ensure that the expensive fast reactors are built in sufficient numbers to transmute the TRU produced in the thermal reactors. This mechanism could take the form of a tax (for example, a charge on LWR electricity or spent fuel) or a legal requirement (for example, a law requiring that all spent fuel undergo separation and transmutation). When the LWR and the burner reactor are in different countries, these mechanisms would have to be incorporated in an international agreement to ensure that the extra costs associated with separation and transmutation are borne by one party or the other or shared between them. If fast reactors prove to be cheaper than thermal reactors, a mechanism would be needed to limit the spread of the cheaper fast-reactor technology and its associated reprocessing and the fabrication of plutonium fuels. It is not realistic to expect current supplier states to retain a monopoly on a reactor technology that generates cheaper electricity, and to expect all other

[27] Assumes spent fuel with an average burn-up of 50 $MW_t d/kg$ containing 1.3 percent TRU (1.16 percent plutonium, 0.06 percent neptunium, 0.06 percent americium, 0.008 percent other), a net thermal efficiency of 33 percent, and a capacity factor of 85 percent.

[28] For an equilibrium system in which all TRU produced by LWRs and fast reactors is consumed by fast reactors, the fast-reactor fraction of the installed nuclear capacity is approximately R/(R–CR), where CR is the conversion ratio and R is the ratio of rate of TRU production in LWRs to the gross rate of TRU consumption in fast reactors per unit installed capacity (250/880 = 0.28 for the assumptions given above).

countries to pay the full costs of a more expensive reactor technology. One possibility might be the leasing of long-lifetime sealed-core reactors to other states, with all fuel manufacture and spent fuel management centralized in supplier states.

NON-ECONOMIC ASPECTS OF THE NUCLEAR FUEL CYCLE

The main incentives discussed thus far to help convince states not to pursue their own enrichment and reprocessing are economic. Historically, however, economic considerations have not been the decisive factors for countries that have pursued enrichment or reprocessing technology. Nations such as Argentina, Brazil, China, France, Germany, India, Japan, and South Africa, and some others, developed their own nuclear fuel cycle facilities. Some of these countries, for at least some of the period of nuclear fuel cycle development, also had nuclear weapons programs that yielded (or could have yielded) nuclear explosive devices. But it does not appear that Germany or Japan has sought to develop nuclear weapons since World War II. For Japan, the main driver was a desire for some degree of energy security, considering the nation's scarce energy resources.[29] But this latter consideration has been invoked for Japan to construct facilities for a full, closed fuel cycle: uranium enrichment, nuclear fuel fabrication, spent nuclear fuel storage and reprocessing, and waste storage (pending availability of waste disposal facilities). Current efforts are attempting to reduce the chance that other countries will follow a similar path. Not all countries share this goal, because of the differentiated status among nations, which it reinforces. Therefore, measures must be taken to allay suspicions and concerns of some countries that the nations with fully developed nuclear enterprises are trying to lock in their nuclear technological and market leadership under the guise of strengthening the nonproliferation regime. Offering all states the opportunity to participate in the profits from multinational or international centers could help address this concern about commercial advantage.

It has been noted that no nuclear power reactor has ever had to shut down because of lack of fuel.[30] This suggests that existing market mechanisms have provided reliable nuclear fuel supplies. New mechanisms designed to increase the assurance of supply should "first, do no harm," and take care not to disrupt the existing nuclear fuel market.

There have been events in the past, however, that have created concerns over the reliability of nuclear fuel supply. Several of the early problems that framed the debate arose from management of the enrichment operations of the U.S. Atomic Energy Commission (AEC), which supplied the entire noncommunist world with enrichment during the 1960s and much of the 1970s (van Doren, 1983; Norman, 1986). In 1966, the AEC decided that it would not accept imported uranium as feed for enrichment contracts for U.S. reactors (which represented a large fraction of world uranium demand), thus creating a major shock in the uranium market and depressing uranium prices outside the United States. Then, in 1973, the AEC announced a drastic revision in the contract terms for enrichment, requiring recipients to enter into long-term fixed contracts at least eight years before the initial delivery. This required renegotiation of

---

[29] Japan had a small nuclear weapons program during World War II. More recently, Japanese documents show that for some key participants (including then Defense Minister Nakasone), gaining a nuclear weapons option for the future was an important part of the reason to pursue a civilian plutonium recycling program (Harrison, 1996, p. 122).
[30] India has had to sharply reduce the power output of some of its reactors due to lack of fuel because India is unable to participate in the international market for uranium. This is discussed later in the report.

several of the U.S. nuclear cooperation agreements, long before the end of their stated 30-year or 40-year terms. The following year, having received a rush of orders for such long-term contracts, the AEC suddenly announced that its enrichment capacity was oversubscribed, that it could accept no new orders, and that some of the orders already received would only get conditional commitments to supply, if sufficient capacity was available, rather than firm commitments. The order books remained closed until 1978. These actions substantially undermined the perception of the United States as a reliable supplier, and increased the priority given to European efforts to establish independent enrichment capacities that were already under way; in addition, the Soviet Union began supplying Europe with enrichment for the first time. Brazil, which had an "Atoms for Peace" agreement with the United States and a contract for Westinghouse to build a set of power reactors, further sought the full range of nuclear fuel cycle facilities—it had uranium mining and milling operations (and large uranium resources) and sought enrichment, fuel fabrication, and reprocessing technologies. The United States would not provide these facilities, especially right after the 1974 Indian nuclear test. With this refusal and the decision to close the enrichment order books, Brazil broke off its negotiation of a new nuclear cooperation agreement with the United States and turned to Germany, which agreed to construct the full set of facilities in Brazil and even to transfer the technologies while assuring supply of enriched uranium from Urenco. Brazil at the same time initiated its own undeclared nuclear weapons program. Brazil abandoned its nuclear weapons program in the 1980s. Ultimately, much of the German-Brazilian deal collapsed, and Germany did not transfer enrichment or reprocessing plants to Brazil. Brazil, however, has established a small enrichment plant, after a long and costly effort.

Also in the mid-1970s, the surge in expected reactor orders and a number of other events caused the price of uranium to shoot upward, creating a major scramble to assure adequate supplies at reasonable cost. Westinghouse, which had committed to supply uranium to fuel the reactors that it had sold at low costs, reneged on this commitment (as the cost of fulfillment was expected to be more than the entire company's net worth).[31]

Then, in 1976, the United States changed its policy from encouraging to discouraging reprocessing of spent nuclear fuel. This provoked tense negotiations with recipient states such as Japan, who were just embarking on reprocessing programs. At the same time, the United States also began to impose new nonproliferation requirements as conditions of supply, culminating with the extensive requirements of the Nuclear Nonproliferation Act of 1978 (NNPA), and in particular the requirement to obtain U.S. consent before any U.S.-origin fuel[32] could be reprocessed or enriched. While most U.S. nuclear cooperation agreements already complied with the NNPA, it imposed substantially higher standards for new agreements and commitments; the U.S.-Euratom agreement did not meet the NNPA requirements, and this led to a brief interruption in U.S. supplies to Euratom and a requirement for annual presidential waivers, both of which the Euratom countries resented. These U.S. steps provoked concerns among some recipient states about what changes in U.S. policy might come next.

During the same period, both proliferation concerns and political upheavals were leading to cutoffs of nuclear fuel supply. After India's 1974 nuclear test, using plutonium produced in a

[31] Westinghouse was sued to force it to fulfill the contracts and did settle the suits with monetary payments (New York Times, 1981).

[32] Material and fuel provided by U.S. companies (and even fuel of U.S. design and any fuel irradiated in a reactor based on a U.S. design) is considered U.S.-origin or U.S.-obligated fuel and requires U.S. permission for transfer to another country.

Canadian-designed reactor using heavy water supplied by the United States, both provided on the condition of exclusively peaceful use, Canada suspended its nuclear cooperation with India. The United States continued to supply low-enriched fuel for India's reactors at Tarapur until supplies to nonnuclear weapon states without full-scope safeguards were prohibited by the NNPA. (That provision also ended or prevented the start of U.S. supplies to several other countries which at that time were not parties to the Treaty on the Non-Proliferation of Nuclear Weapons (NPT) and did not have full-scope safeguards.) India warned that a cutoff of U.S. supply would be a breach of the U.S.-India agreement and would relieve India of its peaceful-use obligations on the plutonium produced at Tarapur, but the United States arranged for France to take over supplying the reactors, avoiding a confrontation with India over supply.[33] That arrangement ended when France agreed to the Nuclear Suppliers Group (NSG) consensus requiring full-scope safeguards as a condition of export. India then received fuel for Tarapur from China for a time (until China also joined the NSG), and most recently from Russia (which argued that an NSG provision allowing supplies to deal with safety concerns justified the export). A major part of the current Indian motivation for negotiating a nuclear cooperation deal with the United States, including seeking a waiver of the NSG restraints, is to gain stable access again to world markets for natural and enriched uranium.

Tarapur and the other cases of cutting off states from nuclear fuel supplies outside the NPT without full-scope safeguards were supply disruptions based on nonproliferation concerns, not politically motivated cutoffs. It is widely accepted that nonproliferation is a condition of nuclear reactor fuel supply, and the mechanisms now being proposed to increase assurance of supply would only be available to states that were in compliance with their nonproliferation obligations (though, as discussed elsewhere in this report, there are varying formulations concerning the particular nonproliferation commitments that would be included in the criteria for participation).

One example of a clearly political cutoff of supply occurred after Iran's 1979 revolution. Germany halted construction of the Bushehr reactor, and the United States cut off supply of highly enriched uranium (HEU) fuel for the Tehran Research Reactor, which was forced to shut down for a time. (Argentina later helped Iran convert the reactor to LEU fuel.) Moreover, before the revolution, Iran was a partner in the French Eurodif international enrichment facility. (This facility had a model similar in some respects to the approach Russia is pursuing at Angarsk, with several countries co-owning the plant and receiving guaranteed supplies of fuel from it, but the technology being restricted to the host state.) The shah of Iran had lent France $1 billion in 1974 to build the plant, in return for access to 10 percent of its output. In 1977, Iran paid another $180 million for future enrichment services from Eurodif. After the revolution, Iran, at that time not interested in nuclear power, wanted its money back, and a lawsuit ensued. In 1991, the suit was settled and Iran received $1.6 billion for its investment plus interest. Iran remained, however, an indirect shareholder in Eurodif through a French-Iranian consortium called Sofidif, which owns 25 percent of Eurodif.[34] When the suit was settled, Iran changed its position and demanded delivery of enriched uranium from Eurodif. France maintained that the contract had expired, and that under the 1991 settlement, Iran had no right to enriched uranium from the facility. No fuel was ever delivered, despite Iran still being part owner of the facility. Iran argues that this experience indicates that joint ownership of foreign facilities does not solve

---

[33] For a discussion of the legal specifics of this episode, see, e.g., Milhollin, 1987; van Doren, 1983.

[34] The French government confirms that this is the arrangement. See http://www.ambafrance-us.org/news/briefing/us110407.asp; accessed June 25, 2008.

the problem of assuring fuel supply (Meier, 2006). The recent experience in which Russian fuel supply to Bushehr was delayed for an extended period as disputes over Iran's nuclear program continued also contributed to Iran's perception that foreign fuel supply is unreliable.[35]

Most other interruptions of supply have not been cutoffs of fuel for existing facilities, but failures to follow through on earlier agreements to build new facilities. There have been many of these, from the Soviet decision to end further nuclear supplies to China in 1960 to the French decision not to supply a reprocessing plant to Pakistan in 1976 to several countries' decisions to end nuclear supplies to Iran in the 1990s.[36] Many of these decisions involved proliferation concerns, and many were made in part as a result of U.S. pressure. While they did not involve cutoffs of fuel supply, they probably did contribute to a perception that foreign nuclear supplies were potentially subject to being interrupted.

Concerns such as these led the countries participating in the International Nuclear Fuel Cycle Evaluation (INFCE) in the 1970s to establish a working group on assurances of supply, and the report from that group suggested a set of steps that are still the main focus of discussion of increased assurances, including backup arrangements among major suppliers and an international bank for enriched uranium (IAEA, 1980). In 1975, IAEA led an initiative studying the feasibility of developing one or more regional nuclear fuel cycle centers, concluding that such centers have economic, environmental, and nonproliferation advantages over national facilities (Meckoni et al., 1977). Similarly, the 1978 Nuclear Nonproliferation Act, while imposing new requirements on U.S. nuclear exports that heightened concerns about U.S. reliability as a supplier, declared that it was the policy of the United States to seek international "mechanisms for fuel supply assurances," and to take "such actions as are required to confirm the reliability of the United States in meeting its commitments to supply nuclear reactors and fuel to nations which adhere to effective nonproliferation policies." The act also urged the President to seek international agreements establishing an International Nuclear Fuel Authority (INFA) that would manage an international fuel bank, with responsibility for ensuring fuel supply on reasonable terms in accordance with agreements between the INFA and supplier and recipient states. The act directed the President to "seek to ensure" that INFA fuel supplies would be "available to non-nuclear-weapon states only if such states accept IAEA safeguards on all their peaceful nuclear activities, do not manufacture or otherwise acquire any nuclear explosive

---

[35] Iran has been found to have repeatedly violated its IAEA safeguards agreement, and the United Nations (UN) Security Council has voted to require Iran to suspend all enrichment and reprocessing activities and imposed sanctions when Iran refused to do so. The governments of the United States and Russia only partly agree on how to proceed with Iran: Both supported the UN resolutions and both joined with European countries in offering incentives to Iran to suspend its enrichment program, but Russia favors an approach more focused on engagement and does not support sanctions as extensive as those the United States would prefer. Russia shipped nuclear fuel to Iran for the Bushehr reactor near the end of 2007, on the condition that the spent fuel be returned to Russia. The United States seeks further isolation and complete cutoff of nuclear and dual-use trade with Iran. The questions go somewhat beyond whether Iran was and is still in violation of its safeguards obligations: How should the international community deal with a nation that was in violation but perhaps technically is not now? The potential for a nation to gain uranium enrichment capability, create a stock of enriched uranium while temporarily in compliance with safeguards agreements, and then withdraw from or violate those agreements by using its nuclear energy enterprise to make nuclear explosive devices is the kind of "breakout" scenario that is viewed as a weak point in the NPT.

[36] Among other examples, Russia agreed not to provide an enrichment plant and terminated cooperation on several other nuclear-related technologies, including laser isotope separation; China agreed to terminate several nuclear projects (in return, in part for entry into force of a U.S.-China civil nuclear cooperation agreement); and Ukraine agreed not to provide the turbine for Bushehr (also in return for cooperation from the United States).

device, do not establish any new enrichment or reprocessing facilities under their de facto or de jure control, and place any such existing facilities under effective international auspices and inspection." (The term effective international auspices clearly refers to some form of international control, but was not further defined.) No steps to establish such an INFA have been taken since the passage of this legislation.

As INFCE came to an end, similar concerns led the IAEA Board of Governors to establish the Committee on Assurances of Supply (CAS) in 1980. While CAS discussed a wide range of assurance of supply issues and proposals over 7 years, it did not reach agreement, and lapsed into abeyance.

More recently, IAEA Director General Mohamed ElBaradei has made efforts to build consensus on ways to renew the nonproliferation regime, halt the growth in nuclear weapons capabilities, lay the groundwork for disarmament, and still enable peaceful use of nuclear energy to expand. In 2003, Dr. ElBaradei offered a three-part proposal: process direct-use material only in facilities under multinational control, with assurance of supply for legitimate use; deploy nuclear energy systems that are designed to avoid creating or using direct-use materials; and consider multinational approaches to the management and disposal of spent fuel and radioactive waste (ElBaradei, 2003). This was followed by the experts' report on multilateral approaches to the nuclear fuel cycle (IAEA, 2005b), and by a working group that reported out in June 2007.

Although not an international center, currently, Brazilian and Argentinean governments are negotiating to establish a binational company for enrichment, without sharing the respective gas centrifuge and diffusion national isotope separation technologies, based on the Eurodif business model, according to a public joint statement from Presidents Lula and Cristina Kirchner in Buenos Aires some months ago (Lasky-Fink, 2008).

Finding 3a

It is feasible to establish a multinational center to provide enrichment services without sharing enrichment technology for countries willing to refrain from developing their own enrichment facility as long as they participate in the center.[37] The International Uranium Enrichment Center (IUEC) in Angarsk, Russia, is one such center. There have been proposals to establish centers under international organizations, although their feasibility has yet to be established. An international dialog, in which concerned countries evaluate the pros and cons of supplementing multinational centers with a center under international control, is needed. Two European multinational consortia have provided enrichment services for two decades: Eurodif, like the IUEC, does not share its technology among its members, but participants need not forgo development of enrichment technology as a condition of participation. Urenco has only three partners, all of which have access to its technology.

A possible way to expand enrichment and reprocessing industrial capacity to cope with a "nuclear renaissance" would be to establish multinational regional large-scale plants, offering to newcomer countries the opportunity to share in their business, administrative, and operative control, but avoiding the spread of technological knowledge that gives rise to indigenous small-

[37] By a *multinational* center, the joint committees mean a facility whose ownership and management involves an arrangement among several countries. Eurodif, Urenco, and the International Uranium Enrichment Center at Angarsk are examples. By an *international* facility, the joint committees mean a facility whose ownership and management is centered in a fully international organization such as the IAEA.

scale plants. These centers could be led by regional powers, such as Brazil and Argentina in South America, and located outside of nuclear weapon states.

Finding 3b
If global usage of nuclear energy increases, it may become increasingly difficult to maintain a system in which nationally controlled facilities in only a few countries provide all enrichment and reprocessing services, as desirable as that might be from a nonproliferation perspective. Offering the opportunity to profit from these technologies may reduce the likelihood that countries would perceive efforts to inhibit expansion of access to the technology as unfair.

As has been seen in the Angarsk facility, setting up an international or multinational center requires addressing many issues, including who owns the technology, what legal arrangements are needed between the center and the customers, and how to obtain competent management of the facility.

Recommendation 3
Over time, Russia, the United States, and other nations should work to create a global system featuring a small number of centers for the sensitive steps of the fuel cycle (especially enrichment and spent fuel management, possibly including storage, reprocessing, or disposal), owned, operated, and controlled by consortia of states or international organizations (but without spreading the relevant technologies beyond existing technology holders). Such a global system, offering many countries the opportunity to participate and share in the profits, would provide a somewhat more equitable and sustainable long-term basis for limiting enrichment and reprocessing facilities to a small number of countries. There has been some criticism that the proposed mechanisms are unfair. The preliminary arrangements should be improved over time.

Reactor fuel is not all the same. Different reactors have different requirements in size and enrichment. With more than 400 reactors in operation, it would not be possible to standardize fuel in a fuel bank. Given that the purpose of fuel assurance is to convince new entrants into nuclear power not to develop indigenous fuel supply facilities but to accept assurance from fuel countries, it would not be wise to attempt to insist that these new entrants use only the fuel supplies endorsed by an international standard. However, it will be interpreted for new entrants to accept the IAEA's safety standards and inspections.

A2. What are the advantages and disadvantages (if any) of establishing international centers for sending and receiving fuel, training personnel, and manufacturing fuel?

INTERNATIONAL MANAGEMENT OF SPENT FUEL

A major reason for these centers is to offer a clear method of fuel assurance, which is a key issue for nonnuclear countries that wish to develop nuclear power. As noted elsewhere in this report, such centers would also enable the nonnuclear countries to share in the profits from enrichment without controlling the technology. The main advantages of centers for providing nuclear fuel and taking back spent nuclear fuel are the incentives that they provide to induce other countries to (a) not develop their own enrichment capabilities and (b) not store and reprocess (or even retain in long-term storage) their spent nuclear fuel. These incentives may come in the form of economic advantage (if the fuel is less expensive), risk reduction (reliability or assurance of supply), and/or unique service offerings (take-back of spent fuel).

Enrichment is a key part of the front end of the fuel cycle. The back end can be "once-through," requiring long-term or permanent storage or permanent disposal of spent nuclear fuel; or it can be part of a "closed" fuel cycle including reprocessing (or regeneration) of spent nuclear fuel, with separation of uranium and plutonium for further use and permanent disposal of waste (see, e.g., NRC, 2003). The current fuel assurance discussions focus on the front end.

Although assuring a supply of fuel is important for a country that is thinking of starting a nuclear power program, perhaps the most powerful incentive to encourage not building an enrichment facility is the offer to take away any fuel supplied after it is spent, either to the country where it was fabricated or to another country. The many proposed fuel assurance arrangements,[38] the virtual or real fuel bank proposed by IAEA Director General Mohamed ElBaradei, the Angarsk center being set up in Russia following the program of former President Vladimir V. Putin, the six-country proposal to the IAEA, the recent German proposal, and the U.S. GNEP program, would use somewhat different approaches to providing an assurance of supply of nuclear fuel. However, none of these programs proposes to take back the spent fuel, although in the past Russia has proposed a program to take back spent fuel and reprocess it or store the waste temporarily.[39] Russia has passed multiple legislative drafts on allowing take-back of Russian spent fuel or import of foreign spent nuclear fuel on certain conditions.[40]

Fuel leasing and other approaches that allowed countries to ship away their spent fuel and nuclear waste (which need not necessarily involve shipping them to the country that provided them) could create strong incentives for countries to rely on international fuel supply, as they would not have to provide a geologic waste repository of their own. Moreover, the availability of such services would greatly reduce countries' incentives to pursue their own reprocessing plants, increasing the chance of limiting reprocessing to a small number of facilities worldwide.

International or regional spent fuel storage and disposal sites could offer other important advantages. Countries might be able to reduce their nuclear waste management costs substantially by pooling their resources to develop a shared repository rather than developing many small repositories. Such facilities could also provide a ready location that could receive irradiated fuels and other materials removed from sites where they pose significant proliferation, security, or safety risks. Discussions about removing plutonium-bearing spent fuel from North Korea, and where it might be shipped, provide a current example of this problem. There are many issues regarding development of regional centers, including whether the international community would support regional centers in the Middle East obtaining sensitive technology.[41]

---

[38] Descriptions included in IAEA, 2007d.

[39] A 2003 workshop by the U.S. National Academies and the Russian Academy of Sciences discussed possible storage sites in Russia at Zhelenogorsk and Krasnokamensk (NRC, 2005).

[40] In the Russian Federation, as a support mechanism for the licensing of nuclear fuel, a special legal amendment/change was made. Currently, nuclear fuel of Russian origin, used in foreign nuclear power reactors, may be sent back to Russia for utilization (long-term, controlled storage, reprocessing, or burial) without return of radioactive wastes and materials. This allows countries that intend to use atomic energy to do so without developing the fuel cycle.

[41] For a discussion of the issues, see Shaker, 2007.

Most countries, however, face substantial political obstacles to accepting other countries' spent fuel and nuclear waste—sometimes portrayed as becoming a nuclear dumping ground. A counter argument can be made if the take-back country views accepting the spent nuclear fuel as advantageous for reasons such as profit or access to the energy resources represented by the uranium and the plutonium separated from the spent fuel for its own future purposes. The waste generated from reprocessing this fuel may be either returned, under contractual terms, to a fuel country exporter in volumes considerably less than those of the spent fuel, or left in a fuel-reprocessing country under certain conditions. To participate, such countries would need to possess modern facilities and technologies for spent fuel long-term storage, innovative reprocessing technologies, and further use of uranium and plutonium in innovative fast-neutron reactors, along with technologies of radioactive waste minimization and safe disposal. Unless reprocessing results in a net improvement from nonproliferation, radioactive waste storage, and economic perspectives, there is no technical or inherent logical reason why a take-back country would need to reprocess the fuel it accepts, so it does not make sense for a take-back program to be the basis for a decision to reprocess. At this writing, it does not appear that a new reprocessing facility would achieve such net improvements.

Russia is the only country now providing fuel leasing (for example, for Bulgaria and Ukraine). None of the fuel assurance programs discuss the possibility of taking back the spent fuel to encourage a new country not to build nuclear fuel reprocessing plants, despite the nonproliferation and potential economic advantages of centralizing them in a few countries. Unfortunately, there is little likelihood that the United States or most other countries will be able to overcome the political obstacles to taking foreign spent fuel in the next few decades. This gradually may become a difficulty in maintaining credibility of programs such as GNEP, which describe fuel take-back as a key element of their approach. At the same time, however, discussions of potential regional or international nuclear waste repositories continue, and over the next few decades, it is plausible that one or more such facilities can be established. This would avoid every country with even one nuclear reactor having to build its own nuclear waste repository.[42]

Even countries that intend to reprocess fuel in a future fuel cycle need not reprocess the fuel until there is a near-term efficient use for the separated constituents. There is, in fact, good reason not to accumulate separated fissile material (uranium, plutonium, and so on) that must be stored and heavily safeguarded until it is utilized sometime in the future, also considering that these materials' characteristics change with time.

Fuel service centers could include storage of natural uranium, uranium enrichment, fuel fabrication, fuel monitoring, and management of spent fuel and radioactive waste. Fuel leasing would offer all of these services and need not ever transfer ownership.

Some people have suggested the idea of nuclear energy "parks" or islands in which multiple nuclear fuel cycle facilities (fuel fabrication, reactors, reprocessing, and even a storage or disposal repository) would be co-located (Ansolabehere et al., 2003, p. 89). This offers some safeguards and nonproliferation advantages—fissile material need never leave the energy park—although the leakage of knowledge and technology is still a concern.

---

[42] For a discussion of the issues surrounding international centers for storage or disposal of spent nuclear fuel, see Bunn et al., 2001, pp. 57-85. A range of more recent documents and analyses related to options for international disposal facilities can be found at the website of the Association for Regional and International Underground Storage, http://www.arius.org; accessed November 26, 2008.

INTERNATIONAL CENTERS AND THE RISK OF TECHNOLOGY LEAKAGE

The chief disadvantage of such centers is a hidden risk that they entail: the potential for leakage of sensitive technology. The leakage of sensitive technology that has probably done the most damage to the nonproliferation regime occurred when A. Q. Khan, working as a contractor on research and development for Urenco, was able to acquire enough information and contacts to build the supply line for Pakistan's nuclear weapons program. Khan went on to form a global black-market supply network that fed into weapons programs in Libya, North Korea, and Iran. Khan's position working on research and development gave him access to many different elements of the centrifuge enrichment technology—access that would have been much more difficult if he were only working at a plant where already-built centrifuges were installed, which is all that is proposed for international staff in most concepts for internationally operated enrichment centers. Nonetheless, if the goal of an international nuclear fuel center is to strengthen efforts to contain these technologies, special efforts are needed to ensure that effective technology controls are maintained, so that the centers do not themselves become proliferators of sensitive technologies.

Different multinational or international fuel cycle facilities and proposals have taken different approaches to this problem. At Eurodif and the Angarsk center, the host state is the only participant with access to the technology, and all of the operating staff for the facility are provided by the host state. In Urenco, by contrast, the partners all have access to the centrifuge technologies used, but are committed by the treaty that established the organization to provide appropriate security for the technology and not to provide it to others. The Urenco consortium has drastically improved its controls over sensitive technology since the Khan episode. Current proposals include the possibility of international enrichment centers with an international staff, but with the centrifuges in "black boxes," so that the staff would have no access to the technology of the centrifuges themselves (Forden and Thompson, 2006).

Such "black-box" arrangements are already being implemented—for proprietary, rather than nonproliferation reasons—for planned U.S. and French enrichment plants that will use Urenco centrifuges, so that the plant staff has no access to the technology. In addition to such physical arrangements to limit access to sensitive technology, it would also be important to establish agreed procedures for security clearances for those personnel who were to be granted access to sensitive information. Similar issues have arisen in international organizations before: for example, when the IAEA was called upon to carry out inspections in Iraq and South Africa that involved actual nuclear weapon design information, the inspectors for that purpose were from nuclear weapon states, with appropriate security clearances from those states.

Detailed review of the specific arrangements for such a facility would be required to ensure that sensitive technologies would in fact be adequately protected, and that the arrangements did not, for example, allow the staff to learn so much about efficient operation of centrifuge cascades that their knowledge might contribute substantially to a nuclear weapons program when they returned to their home countries.

INTERNATIONAL CENTERS FOR TRAINING NUCLEAR PERSONNEL

Large numbers of well-trained personnel will be needed to support the safe, secure, and proliferation-resistant growth of nuclear energy around the world. This will require a major effort in recruiting, training, and retaining people with the required skills, particularly as a large fraction of the existing nuclear workforce is nearing retirement. Currently most of the relevant training is taking place on a national basis, but there may be a role for international training centers in the future. Already the World Nuclear University is providing intensive short courses on key issues to nuclear personnel from around the world,[43] and the IAEA supports a wide range of training courses to help build states' capacity. France, Russia, the United States, and other countries have long provided training for nuclear personnel from other countries. Russian officials have suggested that training centers may be among the network of international centers they envision establishing.

The key advantage of establishing international training centers would be the opportunity to provide consistent education to a wide range of personnel from all over the world, fostering the exchange of ideas and best practices. A key disadvantage, which would have to be carefully controlled, would be the potential for leakage of sensitive knowledge. In particular, personnel from states considering nuclear weapons programs may be able to build networks of personal contacts in the course of training that could allow them to acquire sensitive information and technology that was never intended to be part of the initial training—just as Pakistani and Iraqi scientists exploited such personal contacts to acquire centrifuge technologies in the 1970s and 1980s. Limiting the curriculum of international training centers to nonsensitive topics and ensuring that rigorous counterintelligence programs are put in place could reduce such risks to low levels. Other disadvantages of international training centers compared to national training programs would include the greater difficulty of adapting the training to different national languages, cultures, and circumstances, and the need for students to travel to the location of the international center, possibly for years at a time.

Such training centers might be separate from international fuel cycle centers themselves, to avoid leakage of the sensitive technology used at the fuel cycle center. For the same reason, training personnel on the technical work of fuel cycle services, while important, is not properly the primary role of an international nuclear fuel center. In addition to the technically trained people needed for all fuel cycle facilities, international centers will need people with knowledge and skills in the legal and international relations aspects of the centers' work. This requires some knowledge of the technical side, but particularly required language training, international contacts and training in law. Indeed, there is a need for well-educated, experienced, and motivated professionals. How could this be practically implemented?

Education. At present in Russia, it is difficult to educate students for subsequent work on the nuclear fuel cycle. Although interest in nuclear careers in general is on the rise, relatively few students today plan a career in this area, and some of those who are interested in the field are not aware of their career options and potential opportunities. Therefore, to meet the needs of the international center in Russia, the center participants anticipate that interagency and intergovernmental efforts will be needed to identify and support such students, and coordination will be needed to avoid duplication of specialists' training. Russia has concluded that in this initial stage, it needs to develop a special program utilizing several schools of higher education to train specialists who provide different kinds of support for the center. A system of onsite internships is one organizational approach that may prove effective.

Experience. Before such a program commences and begins to supply the center with specialists, technical support will be provided primarily by those currently working at the center as well as by specialists outsourced to the center on a contractual basis. When selecting these

[43] For a description of the World Nuclear University, see http://www.world-nuclear-university.org.

specialists, those with relevant experience will have a natural advantage; every nuclear fuel cycle plant has its own industrial training center for professional education and retraining of specialists. However, as far as an international center is concerned, additional skills will be needed. These could be met by complementing training in operation with foreign language training and guest lectures by international experts in relevant fields. A system of certificates issued for the center's alumni could be developed. Possibilities of hosting international conferences (under the auspices of the IAEA and other organizations), and the transparent provision of information would also be beneficial.

Motivation. The professional motivation for the center's local staff could be fostered by the provision of a stable and favorable working environment. Incoming professionals, however, may have short-term or long-term employment contracts. Accommodations and educational facilities may be provided for those with short-term contracts or those who are at the center on short-term business or for training. Professionals recruited from ministries, agencies, firms, universities, the RAS, and so on, to work at the center on a long-term contractual basis may be provided with accommodations, a competitive level of compensation, career opportunities, and other benefits. These incentives would require a certain investment from government or private funds or both including joint ventures.

Along with training of technical personnel, professional education in law, international relations, economics, and social and cultural issues is a major aspect of the Angarsk international nuclear fuel cycle center's activities. Therefore, a new institute for nuclear industry training, the International Institute for Energy Policy and Diplomacy, has been established at the Moscow State Institute for International Relations.

One of the institute's key objectives is to train professionals to be able to solve problems of energy policy, diplomacy, and international energy cooperation. For this purpose, the following majors are currently offered to students: international commercial law, world economics, finance and banking, management, and public relations. In 2003, the first graduates received their diplomas, and in 2008 the first master's degrees were granted. The institute facilitates advanced learning of foreign languages. There is a postgraduate program, and the institute issues continuing education certificates. In order to improve education in a core major, the Department of International Energy Sector Studies has also been established at the institute. The programs offer the following core courses: Energy Diplomacy, World Energy, Markets, the Energy Policy of Foreign Countries, Energy Sector Project Management, Environment and Safety in the Energy Sector, the Economy of Energy-producing Countries and Regions, International Energy Security, the Fundamentals of Energy Sector Organization, Management and Economics, Energy Sector Corporate Competitive Strategies, and others.

The institute will grant M.A.'s and Ph.D.'s as well as continuing education certificates. People with a technology background usually are not very aware of legal issues or the social science aspects of international facilities, especially in the field of applicable international law. There are programs to produce technology specialists, but language training and education in legal, economic, social, and cultural issues will be essential to making international fuel centers work.

The world also needs more safeguards, nonproliferation, and security experts. For example, the Moscow Engineering Physics Institute has a graduate program in this area. In Russia, this program is funded by the U.S. Department of Energy's Materials Protection, Control, and Accounting Program, and some U.S. engineering departments are beginning to incorporate courses on these subjects in their curricula. The United States, Russia, and other countries also have centers that provide training in particular tasks related to physical protection, material control, or material accounting, such as the Department of Energy's Safeguards and Security Central Training Academy or the Russian Methodological and Training Center at Obninsk. But there are few opportunities for training that integrate technical matters with a broader nonproliferation perspective. There are not enough people around the world who have an understanding of these issues. The need for these personnel, however, is independent of having centers: Expanding nuclear power around the world will require a lot of expertise.

The Russian institute is an example of how the increased need for personnel to support the fuel cycle can be met. The long stasis in the U.S. demand for new civilian nuclear power reactors led to a greatly reduced interest among undergraduates for nuclear science and engineering programs. This put great pressure on U.S. universities to scale back in these areas. Many did, so that today there are many fewer degree programs available: "[T]he number of university nuclear engineering departments has decreased from 66 in the early 1980's to 30 today." However, undergraduate enrollment has increased "from a low of about 500 in 1999 to over 1,900 in 2007 (APS, 2008)." Therefore, in the United States, nuclear engineering department enrollment is increasing as job opportunities for developing new plants seem to be more attractive than those for maintaining existing plants.

Finding 4
As use of nuclear power grows, there is a need worldwide for well-educated personnel to support the whole nuclear fuel cycle.

Recommendation 4
Countries with large nuclear power programs, such as the United States and Russia, should encourage young people to enter nuclear engineering and related fields and programs that give the breadth of perspective needed.

Finding 5
Arrangements that would provide assured return of spent nuclear fuel could provide a much more powerful incentive for countries to rely on international nuclear fuel supply than would assured supply of fresh fuel, because assured take-back could mean that countries would not need to incur the cost and uncertainty of trying to establish their own repositories for spent nuclear fuel or nuclear waste. Further, it would reduce the number of countries where plutonium-bearing material is stored around the world. Fuel leasing, reactor leasing, and similar approaches could have this benefit, if managed appropriately. For many countries, however, the political barriers to taking back other countries' spent nuclear fuel or nuclear waste are substantial.

Recommendation 5
The United States, Russia, and other suppliers should increase their emphasis on establishing mechanisms for assured fuel-leasing or reactor-leasing services,[44] including take-back of all irradiated fuel. Russia already has legislation and arrangements in place to offer fuel leasing and has such a contract in place with Iran. In both international fuel supply approaches and take-back of spent fuel, Russia is further along in offering services

[44] Today the only discussions of reactor leasing are those on the floating power plants being built by Russia and the nuclear battery being proposed by Toshiba. There will be many legal issues to work out in both cases.

to other countries. The United States and Russia should work together on cooperative approaches that would make it possible to enter into fuel-leasing arrangements in which they would guarantee to supply, and to take back, fuel for the lifetime of reactors built in "newcomer" states, with the fuel taken back to Russia for now, or to the United States as well if circumstances someday make that possible.

Finding 6

A hidden danger of creating such centers is the potential for leakage of sensitive technology. The most damaging leakage of sensitive technology occurred when A. Q. Khan, working as a contractor for Urenco, was able to acquire enough information and contacts to build the supply line for Pakistan's nuclear weapons program. Khan went on to form a supply network that fed into weapons programs in Libya, North Korea, and Iran. An event like this puts the nonproliferation regime in great danger.

Recommendation 6a

The United States and Russia should work diligently with other nations to ensure that all efforts to establish international centers for enrichment, reprocessing, or other sensitive activities include specific, stringent plans to prevent leakage of sensitive information and technology. Plants with staff from countries that do not have technology of the type used at that plant should maintain the sensitive technology in "black boxes" so that the international staff does not have access to the technologies themselves. Plans to prevent technology leakage should be subject to review by a small group of international experts familiar with such technology controls before the centers are established.

Recommendation 6b

Russia, the United States, and other countries working to develop centers should have criteria for participation. Two major criteria for participation by countries beyond the technology holders who provide the technology for the center should be that they not have or be developing an enrichment facility, and that they should be in compliance with IAEA safeguards and nonproliferation obligations.

A3. How should ownership of the nuclear material and the fuel in such arrangements be structured?

There are currently a number of options for fissionable material ownership during the entire fuel cycle. Fuel leasing is the clearest ownership situation: In a lease arrangement, the country of origin maintains ownership of the fuel. Such arrangements are not common. In Soviet times, the arrangement was true leasing: Customers did not own the fuel, even when it was inserted in their reactors; the customers just paid for services. The United States, too, had fuel leasing in earlier years: The U.S. government procured uranium, enriched it, and fabricated fuel, which was then leased to a utility that paid for the energy extracted. This arrangement ended when the law was changed to permit private ownership. An interesting alternative approach is leasing of self-contained, portable reactors, including the fuel in their long-life cores. Russia is building a prototype of a small floating power plant based on icebreaker reactor designs, which it hopes to market for export, and Toshiba is marketing a small reactor that would come with a sealed core containing fuel built in for the entire reactor lifetime. Small sealed-core reactors with lifetime cores and passive safety features could have significant nonproliferation advantages, as could reactor-leasing concepts, and deserve additional research and development to determine if their costs can be reduced (see below for a discussion of advanced technology).

In the regular (nonleasing) uranium and nuclear fuel market today, ownership of (legal title to) a defined lot of uranium is somewhat fungible, like ownership is for other commodities. Ownership can be transferred at each step or process of the fuel cycle, and the end user may simply buy fuel from the fuel manufacturer. Alternatively, some reactor owners buy uranium, retain ownership, and just buy enrichment and fuel fabrication as services. Arrangements exist among yellowcake suppliers,[45] conversion facilities, enrichers, and fuel fabricators for accounting of uranium products and services (most importantly, separative work units, or SWU). An end user may purchase 2 metric tons of yellowcake from company A, purchase conversion and enrichment services from company B, and purchase fuel fabrication services to procure a fuel assembly from company C. The companies will transfer material from A to B and from B to C, but the enriched uranium product transferred from B to C in the end user's name is not typically from the lot of yellowcake that the end user purchased. Likewise, the fuel fabricator often uses enriched uranium from its inventory to fabricate fuel rather than linking a particular delivery of enriched uranium product to the actual fabrication of a client's fuel. The client owns material at any given point, but the accounting of equivalencies in uranium products and services mentioned above facilitates the nuclear fuel supply chain by making the commodity fungible. Ownership of the fuel is transferred either at the beginning or end of shipment of the fuel from the fuel manufacturer to the reactor operator.

The end user may own the fuel in typical nuclear fuel contracts, but provisions, restrictions, and guarantees may be attached that limit the owner's rights to use the fuel in some ways. There are several current examples of such restrictions. As noted above, material and fuel provided by U.S. companies (and even fuel of U.S. design and any fuel irradiated in a reactor based on a U.S. design) is considered U.S.-origin or U.S.-obligated fuel and requires U.S. permission for transfer to another country. The members of the Nuclear Suppliers Group have agreed to impose specific requirements on such retransfers. The Rosenergoatom-Iran contract is a sale, not a lease, but there is a contract guarantee that Iran will return the fuel after irradiation in its reactor. The protections are in place to assure supply can be made by contract or, as in the international centers, under intergovernmental agreements, which are presumably harder to breach than contracts.

For an internationally controlled center supplying fuel to several recipients who were part owners of the center, the fuel might be owned by the international group that owned the center (including, but not limited to, a particular recipient). Another option is that fuel, fuel material, or byproducts within a plant might be owned by particular participants (just as Japanese utilities retain ownership of spent fuel sent to La Hague, as well as the separated plutonium and the vitrified waste from processing in France). Currently enrichment tails belong to the enricher, but this may be revisited in future arrangements because of their value—tails can be further stripped of uranium-235, which is economic for some tails at current uranium prices.[46]

Some issues related to ownership may be important to the success of an approach or scheme (for example, liability and intellectual property), but the specifics of the ownership arrangements may not have a major effect on the nonproliferation issues central to the

---

[45] Yellowcake is a uranium oxide (U_3O_8) produced when a mill separates uranium from uranium ore.

[46] Tails from uranium enrichment are not a proliferation hazard; with enough enrichment capacity, however, the very low concentration could be made into HEU.

committee's concerns. Ownership *per se* is not considered an important proliferation issue as long as full-scope IAEA safeguards are in place and security provisions are adequate to assure that no unauthorized transfers occur. On the other hand, if there are no safeguards and security then it is considered a proliferation threat, regardless of ownership.

This does not mean that there are no ownership issues with multinational or international fuel centers. There are and will be issues of liability, issues of funding because of differences in how different entities handle the cost of time and cost of money, issues of performance, issues of responsibility, and so forth. All of these issues must be resolved between the supplier and the buyer or lessee, but they are not issues of proliferation as long as safeguards and security are in place.

The exceptions are cases in which an ownership arrangement imposes restrictions or commitments that undermine the incentives that are the goal. For example, if transfer restrictions in the context of a fuel bank required the recipient to give up rights to enrich (or to forgo enrichment for a long time), the requirement may be viewed as a too-great commitment for the benefit received and therefore dissuade nations from participating.

Under the IAEA's 2007 proposal, any state that suffered a politically motivated interruption of fuel supply and was in good standing with its nonproliferation and safeguards obligations would be able to draw on the assurances of fuel supply. Under the proposal from the major supplier states, states would only be able to draw on the assurance of supply if they were NPT parties and did not have operational enrichment or reprocessing plants at the moment when they needed fuel. If an IAEA-controlled bank is established on the terms described in the IAEA's proposal, states that supply the material for the bank may still insist that states receiving their material meet additional criteria—such as not having enrichment and reprocessing facilities—before they receive fuel from that particular supplier. It is possible that multiple reserves will be established with somewhat different criteria for drawing on them. Indeed, this already appears to be occurring, as the United States is establishing an LEU reserve on U.S. territory, which will presumably only provide fuel to states that meet U.S. criteria (U.S. Mission to International Organizations in Vienna, undated); Russia is establishing an LEU reserve at the Angarsk enrichment center that the IAEA will be able to draw on to provide LEU to countries suffering a politically motivated interruption; and the IAEA is seeking to establish an additional reserve not on the territory of any current supplier state, to provide additional confidence (using funds from U.S. investor Warren Buffett, the U.S. government, Norway, and other contributors). None of the current fuel bank proposals call for states to give up their right to enrichment and reprocessing forever, or even for a long period, such as 10 years, as such a requirement might be viewed as too great a commitment for the benefit received. Limiting the assured fuel supply to states that do not currently operate enrichment and reprocessing plants would provide an additional incentive for states not to invest in such plants of their own. On the other hand, states such as Brazil and South Africa have strongly objected to such an arrangement, seeing it as an infringement on rights to the peaceful use of nuclear energy under Article IV of the NPT, and since the IAEA Board of Governors operates by consensus, this objection might make it difficult to establish an IAEA-controlled fuel bank with such a requirement. Even a fuel bank without such a limitation would provide additional assurance that fuel would always be available and hence undermine states' incentives to invest in building their own enrichment or reprocessing facilities, and there may be some value in making such assurances available to states such as Brazil, to encourage them to rely increasingly on international supply, and possibly to phase out their own facilities, or at least not increase them to commercial scale.

A4. Should the international facilities be owned by governments or could private companies own some or all of the facilities?

Ownership can be by governments or private entities. In either case, a governmental agreement is likely to be required to establish the legal framework for the international centers, and effective regulation by the host state will be essential. There are examples of international facilities that are government owned, and international facilities that are privately owned. CERN,[47] the international particle physics research center on the Swiss-French border, is fully owned and operated by an international organization with many governments participating. International telescopes in Chile are run by a private company under contract. Both arrangements can work provided that care is taken in establishing and operating them.

VARIANTS ON MULTINATIONAL AND INTERNATIONAL OWNERSHIP AND CONTROL

In addition to the question of private versus government ownership, there are many potential variations on concepts for multinational or international ownership and control of fuel cycle facilities. By a *multinational* center, the joint committees mean a facility whose ownership and management involves an arrangement among several countries. Eurodif, Urenco, and the International Uranium Enrichment Center at Angarsk are examples. By an *international* facility, the joint committees mean a facility whose ownership and management is centered in a fully international organization such as the IAEA. Germany has recently proposed, for example, that a new enrichment plant be established under IAEA control (though managed by a commercial firm), on territory a country was willing to set aside as an international zone (IAEA, 2007c). CERN is arguably a fully international facility (though it could also be considered a multinational facility with a particularly large number of nations participating). There are important differences between CERN and a consortium that operates in the commercial market, but CERN provides a precedent of multinational ownership and governance.

Multinational or international fuel cycle centers might have several nonproliferation benefits. As has already been discussed, states may have more confidence that their fuel supply is assured if they are part owners of such a center and have intergovernmental agreements in place prohibiting any political interference with deliveries. The opportunity to participate in the profits from such multinational or international centers may also reduce states' desire to invest in national facilities of their own.

In addition, many argue that if enrichment and reprocessing facilities *are* established in the future in countries that do not have them today, the resulting proliferation risk would be lower if these facilities were owned and staffed under multinational or international auspices. If many countries owned the facility, there would be a higher—though not insuperable—political barrier to the state where the facility was located (the host state) seizing it and using it to produce nuclear weapons material. Moreover, such an approach with international staff working regularly with the host country's key experts might make it more difficult for those experts to be used to establish covert facilities without any sign of such activity being detected. Furthermore,

[47] CERN is the European Organization for Nuclear Research, a center for particle physics research, technology, collaboration, and education founded in 1954. The organization is run by the CERN Council composed of representatives from the 20 member states.

if such an international-facility regime were in place and widely and successfully used, then if a country decided to begin developing and using these sensitive technologies indigenously, that country's motivation for doing so would legitimately be subject to closer scrutiny, focused on whether the real purpose was to develop a nuclear weapons option. On the other hand, approaches involving international staffing would have to be carefully structured to avoid the centers themselves contributing to proliferation of critical knowledge of how to build and operate enrichment or reprocessing facilities (see discussion of international centers and technology controls).

It may be difficult to convince new states establishing such facilities that they should all be under multinational or international control if existing facilities in major nuclear supplier states remain under purely national control (and even, for facilities in nuclear weapon states, exempted from international inspections). Hence, IAEA Director General ElBaradei has argued that the long-term goal "should be to bring the entire fuel cycle, including waste disposal, under multinational control, so that no one country has the exclusive capability to produce the material for nuclear weapons" (ElBaradei, 2008). Shifting away from purely national control of facilities with the capacity to make large quantities of nuclear bomb material may be particularly important if the world moves toward very deep nuclear arms reductions or a prohibition on nuclear weapons.

Vigorous diplomacy and targeted sets of incentives are likely to be needed to convince countries to participate in international centers rather than build their own facilities, or to establish approaches to multinational or international control for new or existing facilities. In principle, existing nationally controlled facilities could be opened to investment and partial ownership, control, and even staffing from other countries without interfering substantially with their existing operations and contracts, in a way that the host countries can control and build confidence in, so there is no need for countries with such facilities to fear that the international community is somehow going to seize control of these plants. Nevertheless, such transitions are unlikely to be simple or easy. It is likely to be many years before anything like Dr. ElBaradei's vision of a universal regime in which *no* country any longer has purely nationally controlled enrichment and reprocessing capabilities could be achieved. How to structure a step-by-step effort that provides benefits to world security and appropriate incentives for participation at each step will be a critical question.

In addition to its potential benefits, multinational or international control of fuel cycle facilities raises important questions and issues. Questions that will need to be resolved for each center include how key decisions are made, what criteria should make states eligible or ineligible to take part, who gets what share of the profits and losses, who bears what share of the liabilities (such as those for accidents and for nuclear wastes the facilities may generate), how sensitive technologies would be controlled, and how technological improvements would be developed. Choices on these issues have already been made for enterprises such as Urenco, Eurodif, and the Angarsk center. Additional choices will have to be made as these enterprises evolve and additional multinational or international centers are established in the future. In general, any center, whether national, multinational, or international, may require a unified management structure, so that key decisions can be made efficiently. Similarly, any center will have to be regulated appropriately; for Eurodif, Urenco, and Angarsk, the host state where the facility is located has always maintained the authority and responsibility to set and enforce appropriate safety, security, and environmental rules, and this is likely to be the case for future facilities as well.

A wide range of multinational or international approaches is possible. Variations along several dimensions are particularly important.

Control of sensitive technology. In some approaches, only the host state has access to the sensitive technology used at the center. This is the approach taken in the Eurodif consortium and the Angarsk center. In such cases, there would in general be no more danger of technology leakage than there is for purely nationally controlled facilities. In Urenco, by contrast, all of the Urenco partners have access to Urenco's centrifuge technologies. In principle, in approaches where one partner provides and controls the technology, that partner need not be the state hosting the international facility. For example, new enrichment plants using Urenco centrifuges are scheduled for construction in both France and the United States, with the centrifuges in "black boxes" that the United States and France have no access to; some analysts have proposed that an enrichment plant with multinational ownership and staffing be established in Iran on a similar "black box" basis (Forden and Thompson, 2006). Clearly, if multinational or international centers are to avoid themselves becoming a source of proliferation of nuclear-weapons-related technologies, plans for how the sensitive technologies used at each center will be controlled will be very important. In general, centers with a variety of states participating should limit access to sensitive technologies to personnel from states that already possess these technologies. (See the discussion of technology controls for international centers.)

Degree of multinational or international sharing of ownership. In some approaches, the partners might have shares of the ownership and control of the facility small enough that no one partner had control, and all major decisions would require support from several countries. In Eurodif, by contrast, France, the host state, has always maintained majority ownership, so that ultimately France can control all of the consortium's key decisions. Similarly, Russia has indicated that it plans to maintain a majority of the shares of the Angarsk enrichment center. In these cases, the minority partners may get little actual control of the center, though they do get to share in its profits. Fully international ownership would presumably mean that the actual equity ownership of the facility would rest in the hands of an international organization, and a large number of states within that organization would have to support each major decision. This would include the financial aspects, so that all profits or losses would be internationally shared, and annual budgets would be approved by a board of directors, presumably appointed by the international organization. (The international organization could be an *ad hoc* organization established solely for this purpose, or could be an existing organization that also exists for other purposes, such as the IAEA.) If an important part of the reason for placing a facility under multinational or international auspices is to increase the international community's confidence that the plant will not be turned to weapons use, each particular arrangement will have to be reviewed to see if the approach to multinational or international control will meet that purpose; a multinational consortium consisting of several allied states perceived to be bent on developing nuclear weapons, for example, would do little to increase international confidence in the peaceful nature of the facility.

National, multinational, or international staffing. As noted above, facilities with a multinational or international staff have both advantages and disadvantages. The advantage is increased transparency in the operations of the facility and the activities of the host country's experts in that technology, making both covert diversion and construction of covert facilities more difficult to accomplish without detection. The disadvantage is the potential for leakage of sensitive technology to participants on the multinational or international staff who are from countries seeking those technologies (if the staff includes individuals who are not from countries that already have all of the relevant technologies). For facilities with key technologies (such as centrifuges) in a "black box," it would be important to understand whether the knowledge and experience of, for example, cascade operations that the multinational staff would gain would still make an important contribution to a weapons program. In principle, it should be possible to design facilities with important parts of the facility staffed by multinational teams that did not spread any critical fuel cycle knowledge.

Along these and other dimensions, centers could be closer to or further from purely national ownership, control, and staffing, and each variation would have somewhat different nonproliferation impacts.

A5. What regulatory requirements should be in place in the receiving country to provide assurance of safety and safeguards?

If a country is to participate as a "recipient country" in any of the fuel-assurance and fuel take-back schemes under discussion, it is highly desirable that the country have in place laws, regulations, and procedures that meet international norms for safeguards, safety, and physical security.[48]

Concerning safeguards, all of the proposals now under consideration include a requirement that the recipient country be in full compliance with its international obligations according to the NPT and the IAEA's safeguards regime. Different countries, however, have entered into different obligations, which may affect their eligibility under different proposed approaches. In some approaches, for example, the recipient country would have to have accepted safeguards on all its nuclear activities, and possibly also the Additional Protocol, to be eligible to participate. In other proposed approaches, countries such as India, which have safeguards on only a portion of their nuclear activities, would also be able to be recipients of assured fuel supplies. At a minimum, the material provided under such an arrangement should itself be under safeguards to assure its peaceful use, and this must be monitored by the national, international, or multinational center that is supplying services.

Concerning safety, if the country is operating one or more nuclear power reactors, it is necessary that an effective nuclear regulatory agency be in place, along with a legal framework (laws, regulations) that provides the wherewithal for that agency to perform its work effectively. While the joint committees do not expect that a supplier of fuel cycle services will be required to monitor this aspect explicitly, or to deny such services on the basis that the country does not have an effective regulatory regime, the committee does expect that there will be enough international

[48] With respect to safeguards, all nonnuclear-weapon states that are parties to the NPT are required to accept IAEA safeguards on all of their civil nuclear activities, and to have state systems of accounting and control of nuclear materials that are comprehensive and accurate enough to serve as the basis for declarations and inspectors' checks of the accuracy of those declarations. (Nuclear weapon states and states not party to the NPT are not required to accept safeguards on all their civil nuclear activities, so they do not have the benefit of this international discipline on the quality of their national nuclear material accounting systems; in some cases their domestic standards for the accuracy of nuclear material accounting are quite different from the IAEA's international standards.) With respect to safety, in most cases recipients would be expected to be participants in the major nuclear safety and nuclear liability conventions, and it is essential that they have an effective nuclear regulatory body in place with the independence, expertise, power, and resources needed to do its job. With respect to physical security, recipients would typically be expected at least to follow the minimal requirements established in the Nuclear Suppliers Group guidelines; many suppliers may call for recipients to meet higher standards, including participation in the Convention on Physical Protection of Nuclear Materials (and its amendment, once it enters into force), and protecting materials and facilities in a manner consistent with IAEA recommendations.

attention paid to this issue to assure that no country can possess and operate nuclear power plants without it. Several mechanisms for assuring this seem likely. For example, it is unlikely that a reactor manufacturer or vendor would sell a reactor to such a country, nor provide services to it; many companies have accepted the reasoning that a nuclear accident anywhere is a disaster for nuclear power everywhere, and would be especially damaging to the particular vendor's business. Also, the IAEA would presumably be alert to the situation, and would help to bring international pressure to bear.

The IAEA, several of the nuclear suppliers (the European Union, France, the United Kingdom, the United States), and a few other countries have for over a decade been coordinating a broad program of unilateral and sometimes multilateral technical, legal, and training assistance to developing countries in properly establishing and operating a nuclear regulatory agency.[49] This program has been quite successful whenever a recipient country has embraced the ideas, which many but not all of them have. The World Association of Nuclear Operators, an industry group, has also played an important role in exchanging best practices and lessons learned and organizing international safety peer reviews.

Appropriate nuclear security is another aspect of the overall safety regime that the country's nuclear regulatory authority would need to assure. Here too, the IAEA, the United States, and a few other countries have been working to assure that each country with weapons-usable nuclear materials or nuclear facilities whose sabotage would lead to serious consequences puts in place appropriate physical protection and material accounting measures, but there is a great deal still to be done, including to convince countries that nuclear theft and sabotage are real threats deserving substantial investment to address.

A6. What level of technical personnel are needed, in terms of training and in terms of numbers, to provide adequate confidence that the countries receiving fuel can safely and securely operate their reactor(s)?

It is not an appropriate role for an international fuel cycle center to ensure this training and confidence. Substantial experience and knowledge about this question exists in the countries with many nuclear plants (France, Japan, Russia, South Korea, the United Kingdom, and the United States) that can address this question. As with Question A5, this is really not a question that is unique to international centers.

A7. What could be the role of the International Atomic Energy Agency in overseeing the transfer, use, and/or return of fuel?

For the new international center at Angarsk, this is a question that was being worked out by the IUEC and the IAEA as this NAS-RAS study was being carried out. The United States and possibly other countries hope that arrangements worked out between the IAEA and the IUEC will fulfill nonproliferation and other goals so that they can serve as a model for other centers and fuel reserves to follow in their arrangements with the IAEA. The bylaws of the IUEC say that the IAEA should have a "major role" in the work of the center and that it will be under IAEA safeguards. In particular, it appears that Russia and the IAEA have tentatively

[49] One example is a program called CONCERT, begun in 1992 by the European Commission, which established nuclear regulatory cooperation and assistance with countries in Eastern Europe (see http://ec.europa.eu/energy/nuclear/safety/programmes_en.htm; accessed November 26, 2008.).

agreed that if an IAEA member state in good standing with its nonproliferation obligations suffers an interruption of nuclear fuel supply that it cannot address by other means, it will be able to make a request of the IAEA, which will then be able to draw on the stocks of LEU stored at the IUEC to fulfill the request. Assuming that this arrangement is, in fact, established, the IUEC will become, in effect, the first international fuel bank—though the IAEA is still working to establish one or more additional fuel banks located outside current nuclear fuel supplier states, to further increase states' confidence that supplies cannot easily be interrupted. In short, for the Russian Federation, the United States, and other countries, the IAEA could serve as an important conduit and buffer between suppliers and recipients in the context of fuel service centers, fuel banks, and other fuel service arrangements.

As with the IUEC, the joint committees expect that any center would have IAEA involvement, especially to fulfill safeguards obligations. Such centers need the IAEA to certify that a country meets safeguards and nonproliferation criteria prior to shipment; to oversee shipments to make sure they meet international standards for physical protection, safeguards, and safety; and to inspect safeguards for the facilities while the fuel is in the recipient country.

A8. What changes in laws and regulations in the countries sending, consuming, and receiving spent fuel would be required to implement an international assured fuel cycle concept?

The internationalization of the fuel cycle will require new laws and regulations in both countries hosting such centers and in countries using the centers. In some cases these will be new laws to address new issues raised by the international or multinational aspects of the agreements. In other cases, laws do exist but they differ from country to country, so that either new laws or revisions to existing laws must be made to arrive at a common basis. In addition there are several different concepts for internationalization, some with varying contract commitments, depending on the needs and desires of the various countries involved. As a result, it is not possible to identify generic changes that apply universally.

Russia has modified several laws to enable the establishment of the IUEC and other international centers. In particular, changes to the law have made it possible for entities other than the Russian government (including foreigners) to be partial owners of nuclear facilities; modified restraints on foreign access to Angarsk; placed Angarsk on the eligible list for safeguards under Russia's voluntary offer agreement with the IAEA; and made it possible to import foreign spent fuel for storage or reprocessing in Russia.

In 2001, the State Duma, the lower house of the Russian parliament, approved a set of bills and amendments to the laws adopted earlier. The functions and objectives of and the enforcement procedures for such laws are as follows.

1. The law permitting import of spent fuel to Russia authorizes import of spent fuel to Russia for reprocessing and long-term storage at controlled sites. Spent fuel generated from nuclear fuel of Russian origin will be imported for the purposes of reprocessing and long-term storage at controlled sites, with an option (assurance) for permanently keeping in Russia all kinds of radioactive waste and fissile materials. Decisions to keep radioactive waste and fissile materials in Russia in cases where the spent fuel is from nuclear fuel of Russian origin will be made by the Russian Federation government in the form of an intergovernmental agreement. Spent fuel from nuclear fuel of foreign origin may also be imported to Russia for reprocessing and long-term storage at controlled sites, provided, however, that the requirement for repatriating

the radioactive waste be given priority. A special committee of members equally representing the Russian government, the State Duma, and the Federation Council will make a decision on importing any spent fuel of foreign origin, and such decision is to be approved by the President of the Russian Federation. Academician Nikolay P. Laverov is currently chairing the committee.

2. The law on dedicated environmental foundations requires that in cases where spent fuel of any origin is imported to Russia, part of the proceeds from such activities be mandatorily redirected to finance specific projects developed for improving the environment in the regions where such activities occur.

The provisions of the laws listed above are currently applied to the import into Russia of spent fuel from the research reactors that were built in eastern European countries (under the Soviet Union) or the Commonwealth of Independent States and then loaded with Russian nuclear fuel. Since 2001, no spent fuel not of Russian origin has been imported.

There are many laws and regulations that would need to be revised to reduce barriers to proliferation threat reduction. One important constraint in U.S. law and policy relates to management of spent fuel that has U.S. obligations attached to it under the Atomic Energy Act (AEA) of 1954, as amended and revised. This includes fuel that was mined, enriched, or fabricated in the United States, or irradiated in a reactor with major components based on U.S. technology. Under the AEA, countries with such fuel may not transfer it to other countries without U.S. permission, and the United States cannot legally give its permission unless it has a civilian nuclear cooperation agreement (known as a 123 agreement, referring to the relevant section of the AEA), with the country where the fuel is to be shipped. Hence, international centers for spent fuel management would not be able to handle U.S.-obligated fuel—representing a substantial portion of the world's stock of spent fuel—unless the United States had a 123 agreement in place with the country where the center was located, and a policy of approving the transfers. The United States and Russia have recently negotiated such an agreement (see Appendix D), but as of 2008 it had not entered into force, and some members of U.S. Congress were arguing for delaying or blocking its implementation. Such an agreement would be necessary for a future international center for spent fuel management to be able to operate effectively in Russia. Politically the United States is unlikely to be able to take back spent fuel itself for many years to come. Under U.S. law, such take-backs would require congressional approval, though they are not prohibited in principle. Such approval is unlikely to be forthcoming, except in special cases such as the ongoing return of irradiated research reactor fuel, which is part of a program to reduce proliferation risks by eliminating HEU from as many research reactors as possible. See Section B6 and Finding 12 for more on this topic.

Finding 7

Safeguard arrangements, fuel transfer processes, and return of spent fuel provisions are only a few of the complex legal issues that must be resolved if fuel assurance, fuel take-back, and multinational or international fuel center programs are to be effective.

Recommendation 7

The IAEA should lead an international effort to identify these legal questions and options to be considered. The IAEA should also convene countries to reach agreement on preferred solutions.

CHAPTER 3

FUEL REGENERATION OPTIONS TO SUPPORT AN INTERNATIONAL NUCLEAR FUEL CYCLE

Primary Issues:
B1. Compare the uranium recovery by extraction plus (UREX+), the plutonium and uranium recovery by extraction (PUREX) process, and other processes being considered by the Russian Federal Agency for Atomic Energy for separation of fissile and other materials from spent or irradiated nuclear fuel. Consider the resulting waste streams and what can and should be done with these waste streams.

B2. Compare the burn up and the number of cycles needed to reach an acceptable level of destruction of actinides in the conceptual advanced burner reactor proposed in the U.S. Global Nuclear Energy Partnership (GNEP) and in the Russian BN-600 and BN-800 reactors.

COMPARING NUCLEAR OPTIONS: THE NEED FOR A SYSTEMS APPROACH

The joint committees believe that a comparison to make choices among different fuel cycle options (reactors, fuel types and sources, spent fuel management, and processing) must use a systems approach. Such analyses would consider the entire life cycle of proposed nuclear energy systems, integrating assessments of fuel processing, fabrication, reactor design, and more. Only in this way can key trade-offs be made among different parts of the system. It is likely that the best technologies for processing spent fuel will be different depending on the specific reactors in which the processed materials will be irradiated, and the fuel fabrication approaches for them. In the U.S. case, for example, the Experimental Breeder Reactor II (EBR-II) Program was successful because fuel fabrication, reactor design, and spent fuel processing were done in an integrated way, making it possible to optimize choices for the system as a whole.

Good decisions among different proposed processing-fabrication-reactor systems require clear, consistent, and well-thought-out criteria, based on justifiable system objectives. Picking a particular numerical target for some system characteristic (such as 99.99 percent purity for uranium separated from spent fuel) without careful analysis of the overall system benefits and costs of meeting that goal leads to poorly optimized systems. Building in assumptions or early decisions, such as a requirement for either a once-through or a closed fuel cycle or a particular reprocessing technology, allows a systems analysis to consider only variants of the already-chosen approach. A good goal would be an integrated reactor fuel cycle system that offers the best combination of economics, safety, security, proliferation resistance, environmental impact, process operability, and sustainability, given the situation that exists for a nation at a particular time.

In many cases some systems may offer more promise on some of these criteria, while others look better with respect to other criteria, making trade-offs inevitable. Whether more emphasis should be given, for example, to saving money or to reducing environmental impact is not a technical decision but one based on values, which must ultimately be made by society, through a political process. The role of designers and technical experts is to make clear the choices and trade-offs that need to be made, outline the benefits and downsides of each of the leading approaches, and do their best to ensure that the decisions ultimately made are well informed and carefully considered.

Criteria for Comparison

Each of the key criteria mentioned above can be specified in more detail, so as to provide more detailed guidance to those designing and assessing these systems.

Economics. Each system can be compared based on its life-cycle electricity cost. Additional criteria may include the degree of uncertainty of those cost estimates; the system's contribution to the costs of spent fuel and nuclear waste management; initial capital costs and the resulting level of financial risk in implementing and operating a system; the variability and reliability of the electrical output; and the system's attractiveness or unattractiveness to the private sector (along with the scope of required government subsidies or regulations needed to make the system competitive).

Safety. Each system can be compared based on the overall risk of a significant accident it poses (including both the probability and the consequences of the various types of plausible accidents in the system); accident reports by regulatory agencies and others can provide insight into risks. Radiation doses to the public and industrial safety during normal operations are also considerations, though these risks are low for most proposed systems. Because the risks of significant accidents may be difficult to estimate rigorously and compare among systems that have never been built, decision makers may choose to focus on the degree to which known risk factors are present and how they are addressed (such as positive coefficients of reactivity, which can result in power excursions), or the degree to which known safety factors are present (such as "passive safety systems").

Security. Thorough security comparisons would examine how difficult it would be for adversaries to cause a major radioactive release through sabotage, or through the theft of material that could be used to make a nuclear device. Systems that continuously maintain the nuclear materials in their cycle in forms that could not be used in weapons without either isotopic enrichment or extensive chemical processing using heavy shielding rank better on this criterion. Reactors with greater degrees of inherent safety and widely separated redundant safety systems so that they would be more difficult to sabotage simultaneously are also more inherently secure, according to this measure.

Proliferation resistance. The proliferation resistance of alternative nuclear systems depends on how difficult it would be for a nation or a subnational group to use a facility or material to make a nuclear explosive device. No chemical processing facility can be constructed to make it impossible to change its product streams, but it can be designed to make changes costly, lengthy, and detectable. Proliferation resistance can be judged by criteria related to the material streams and the processes, including the extent to which (a) access to the material, facilities, and technologies used in the proposed cycle would reduce the time, cost, and observability of producing weapons-usable material;[1] (b) the personnel and experience involved in operating the proposed system would reduce the time and cost to produce weapons-usable material (not only at the facilities in the proposed system but at other, possibly covert, facilities); (c) the difficulty of ensuring against sensitive leakage of technology might increase or decrease if the proposed fuel cycle were implemented; (d) the number of safeguards inspection days per gigawatt-day (GW-d) generated would increase or decrease in the proposed fuel cycle, compared to other systems; and (e) the uncertainty in meeting safeguards goals would increase or decrease for the proposed system compared to other systems. In addition, one needs to consider how the adoption of the proposed system by some countries might affect other countries' decisions to pursue sensitive technologies such as enrichment or reprocessing.[2] With fuel cycle facilities and processes in particular, useful objectives include ensuring that conversion of material from reactor fuel material to directly usable weapons material would be difficult, time consuming, and have a high probability of being detected (see Box 3.1).

A facility that achieves these objectives would have *no* separation or processing facilities that (a) have directly usable material in storage, (b) have directly usable material at any other point in the fuel cycle, (c) offer a way to produce directly usable material by simple process changes, (d) offer a way to produce directly usable material without substantial equipment replacement or major modifications, (e) offer a way to carry out such equipment or plant modifications with facilities and components normally onsite, or (f) offer a way to carry out equipment or plant modifications without plant decontamination or entry into extremely high radiation fields. In addition, such a facility would (g) have uranium-handling equipment for all stages of the fuel cycle that are designed for criticality safety when handling low-enriched uranium (LEU), but not when handling highly enriched uranium (HEU), so as to deter using it for higher enrichments than those for which it was designed; and (h) provide a high likelihood of timely warning—that is, the length of time required after likely detection of a diversion effort and before sufficient material was available for a small nuclear arsenal would be such that there is time for national and international bodies to respond.

Environmental impact. All proposed systems would be expected to meet all applicable environmental, safety, and health requirements. The environmental impacts of a fuel cycle depend sensitively on the details of the fuel cycle and how it is implemented and operated, and it is difficult to argue for holding today's proliferation and other problems at risk for tomorrow's unknown problems. A system can therefore be evaluated based on whether it would significantly increase existing environmental, safety, or health risks beyond those that would exist if it were not implemented. Thorough comparisons among different nuclear systems would be based on expected harms to the public, workers, and the environment throughout the life cycle of the system from both radiation and other industrial or chemical impacts. This would include both normal operations and plausible accident scenarios. Variations among doses of radiation that are all very low may not be particularly important discriminators between one system and another, however.

---

[1] A related metric is how amenable the process is to safeguards, particularly the relative ability to meet International Atomic Energy Agency (IAEA) goals for timely detection of diversion of a significant quantity of weapons-usable material.

[2] For a discussion of similar criteria, see Bunn 2007, and Nuclear Energy Agency for the Generation IV International Forum, 2006.

Resource utilization. Proposed systems can be compared on the basis of how long they could continue to generate electricity economically given likely future system constraints, including the cost of uranium and repository capacity. It will not be urgent to shift toward closed fuel cycle systems that utilize uranium more efficiently until the cost of fuel from fresh uranium persistently exceeds the full cost of fuel from recycled fissile material or other factors, such as constraints on repository capacity, become overriding factors.

Technical feasibility and maturity. Admiral Hyman Rickover pointed out the perils of comparing "academic reactors" and "practical reactors."[3] Comparisons of proposed future systems must take into account their respective levels of technological development, as it is often the case that as work focuses on a specific design, problems arise that were not anticipated at earlier stages of development. Proposed systems can be compared based on the presence or absence of required steps whose technical feasibility is not yet established, on the level at which individual steps and the total system have been designed and demonstrated, and on the estimated years and resources that would be required to prepare the system for commercial deployment.

Advanced safeguards and security technologies could play a critical role in pursuing the nonproliferation goals mentioned above. In particular, in providing increased capabilities to detect covert nuclear facilities; highly accurate near-real-time monitoring of material flows in bulk processing plants with reduced intrusiveness, increasing confidence that any diversion would be detected; low-cost real-time monitoring that would set off an immediate alarm if stored nuclear material were tampered with or removed; effective protection against sophisticated outsider and insider theft and sabotage threats at reduced cost; and design of facilities to simplify and increase the effectiveness of safeguards. A study group of the American Physical Society concluded that a reinvestment in research and development on safeguards and security technologies is needed (APS, 2005), and the joint committees agree.

---

[3] In 1953, in the face of criticism of the U.S. Atomic Energy Commission plan to develop pressurized water reactors rather than exploring the multitude of other reactor options, Admiral Rickover wrote (Rockwell, 2002; Kuliasha and Zucker, 1992, p. 271; and Rickover, 1970, p. 1702):

> An academic reactor or reactor plant almost always has the following basic characteristics: (1) It is simple. (2) It is small. (3) It is cheap. (4) It is light. (5) It can be built very quickly. (6) It is very flexible in purpose. (7) Very little development will be required. It will use off-the-shelf components. (8) The reactor is in the study phase. It is not being built now. On the other hand, a practical reactor can be distinguished by the following characteristics: (1) It is being built now. (2) It is behind schedule. (3) It requires an immense amount of development on apparently trivial items. (4) It is very expensive. (5) It takes a long time to build because of its engineering development problems. (6) It is large. (7) It is heavy. (8) It is complicated.

BOX 3.1
DIRECTLY USABLE MATERIAL

The joint committees use the term "directly usable" to mean that the material could be used to fabricate a nuclear explosive without extensive chemical processing using heavy shielding or isotopic enrichment. As examples, fresh LEU fuel and spent fuel from a typical power reactor would not be directly usable weapons materials by this definition, as LEU would require isotopic enrichment before it could support an explosive nuclear chain reaction, and spent fuel from typical power reactors could only be processed if some form of heavy shielding were used. By this definition, unirradiated mixed-oxide (MOX) or transuranic (TRU) fuel or uranium-aluminum HEU research reactor fuel would be considered directly usable, because, while each would require chemical processing before it could be used in a nuclear explosive, that chemical processing would not have to be done remotely and would pose fewer challenges.* The joint committees' use of directly usable weapons material is very similar to the International Atomic Energy Agency's (IAEA) term "unirradiated direct-use material," which refers to direct-use material (including chemical mixtures such as MOX) "which does not contain substantial amounts of fission products; it would require less time and effort to be converted to components of nuclear explosive devices" than would, for example, plutonium in spent nuclear fuel.

*For a discussion of the relative availability of different types of adversaries to recover material usable in a weapon from different types of materials, see NRC, 2000.

NOTE: For the IAEA definition, see IAEA Safeguards Glossary, accessed at www.pub.iaea.org/MTCD/publications/PDF/nvs-3-cd/Start.pdf on July 19, 2005.

The Generation-IV International Forum (GIF) has outlined an approach that is similar in some respects to the system-level, criteria-based approach advocated here. GIF's "technology roadmap" emphasizes the need to focus on entire nuclear energy systems, including "the nuclear reactor and its energy conversion systems, as well as the necessary facilities for the entire fuel cycle from ore extraction to final waste disposal" (DOE, 2002, pp. 5-6). GIF has specified several ambitious goals for such systems (though it remains unclear whether any single system can meet all of these objectives simultaneously).

Sustainability. The goals are to develop systems that will "provide sustainable energy generation that...promotes long-term availability of systems and effective fuel utilization for worldwide energy production," and "minimize and manage their nuclear waste and notably reduce the long-term stewardship burden, thereby improving protection for the public health and the environment."

Economics. The goal is a system that "will have a clear life-cycle cost advantage over other energy sources," and "a level of financial risk comparable to other energy projects." To complete the economic analysis, a discount rate must be selected and its basis carefully explained.

Safety and reliability. Goals for Generation IV systems are to "excel" in safety and reliability, and in particular to have "a very low likelihood and degree of reactor core damage" and to "eliminate the need for offsite emergency response." (The goal of eliminating all reliance on emergency responses outside the site is an example of setting very specific goals within an overall category, possibly without adequate consideration of the costs and benefits of that particular objective.)

Proliferation resistance and physical protection. GIF set the goal of "increasing the assurance" that these systems would be "a very unattractive and the least desirable route for diversion or theft of weapons-usable materials," and that they would "provide increased physical protection against acts of terrorism." As stated, these are notably less specific than the goals for economics or safety and reliability.

EVALUATING CURRENTLY PROPOSED SYSTEMS

Nations that have led technological development of nuclear fuel cycles, including France, Japan, Russia, the United Kingdom, and the United States, have developed a variety of technological options for processing spent nuclear fuel. Some processes, including the only ones deployed on a large scale, initially were developed and optimized for the military purpose of extracting plutonium for nuclear weapons. Some of those processes have been adapted for nonmilitary applications, specifically for processing different types of commercial nuclear fuels. Each of the processes is actually a family of processes (variants on the overall process approach; no two PUREX lines are identical). The most important of these families are PUREX, COEX, UREX(+), pyroprocessing, fluoride volatility, REPA, TRUEX, and supercritical carbon dioxide (CO_2). Among these, PUREX, COEX, UREX+, and pyroprocessing garner the most attention today in nations with grand nuclear energy ambitions. Box 3.2 gives descriptions of these processing options. The descriptions are necessarily at a high level because many variations within the same family are possible, and two variants can have important differences (see Box 3.3 for an illustration of a processing family, UREX+). One of the reasons why variants exist within a family is that it is necessary to tailor a given process specifically to deal with each different fuel type, or even to deal with very different burn-ups of the same fuel type. For this reason, even with this set narrowed, it is not really possible to carry out a detailed comparison among the options, as described in greater detail below.

BOX 3.2
MAJOR TECHNOLOGICAL OPTIONS FOR PROCESSING SPENT NUCLEAR FUEL

PUREX

The PUREX process coextracts and then individually separates to desired purity uranium and plutonium from fission products and other transuranics. Those transuranics and fission products become part of the waste stream. The plutonium can be used in fabrication of mixed-oxide or metallic fuel. The uranium can be reused, too, but uranium from commercial reactors typically is not reused, because the isotopic mix of irradiated uranium is not optimal and fresh uranium is relatively inexpensive. However, uranium recovered from research and propulsion reactors is sometimes recycled.

COEX

The COEX process is a modified version of PUREX that coextracts roughly equal amounts of uranium and plutonium for fabrication into MOX fuel. Minor actinides go to the high-level waste product along with the remaining fission products.

UREX and UREX+

The UREX process removes uranium in an initial extraction step. That uranium is purified for disposal as low-level waste or for reuse. The remaining stream of transuranic constituents, including plutonium, is maintained as a group and destined for fabrication into fast-reactor fuel. Fission products are a separate stream, but some of them may be separated further (UREX+). For example, in some schemes the plan is to separate cesium and strontium from the other fission products and store them for decay, to reduce repository heat load, which for some repositories may increase effective repository capacity. Lanthanide fission products may be retained with the transuranics if they are deemed to provide some self-protection radiation barrier, or they may be left with the other fission products.

Pyroprocessing

There are different processes that were initially developed in Russia and the United States. Each country is continuing to develop its own approach, and France and Japan are also conducting research on their own approaches.

U.S. process: Spent fuel, if oxide, is reduced to a metallic form and immersed in a bath of molten salt floating on a liquid cadmium cathode, which attracts plutonium and the minor actinides. Uranium can be deposited on a solid cathode. This process, never deployed at any significant scale, would be most readily applied to metallic fuel. The United States also developed a melt-refining process for pyroprocessing a special sodium-bonded fuel from the EBR-II, and ran an extensive processing campaign for several years, but the direct applicability of this process to other types of fuels is probably limited.

Russian process: Spent fuel is dissolved in molten salts and crystal plutonium dioxide or electrolytic plutonium, and uranium dioxides are recovered from the melt on a solid cathode. Uranium and plutonium remain together. This process is most readily applied to oxide fuel.

BOX 3.3
THE UREX+ FAMILY OF PROCESSING OPTIONS SPENT NUCLEAR FUEL

Table Stages and Products From UREX+ Variants*

Process	*1st Product*	*2nd Product*	*3rd Product*	*4th Product*	*5th Product*	*6th Product*	*7th Product*
UREX+1	U (highly purified)	Tc, I (LLFPs, dose issue)	Cs, Sr (short-term heat mgmt.)	Other FPs	TRU+Ln (temporary storage)		
UREX+1a	U (highly purified)	Tc, I (LLFPs, dose issue)	Cs,Sr (short-term heat mgmt.)	FPs (including lanthanides)	TRU (group extraction)		
UREX+2	U (highly purified)	Tc, I (LLFPs, dose issue)	Cs,Sr (short-term heat mgmt.)	Other FPs	Pu+Np (for FR recycle fuel)	Am+Cm+ Ln (temp. storage)	
UREX+3	U (highly purified)	Tc, I	(LLFPs, dose issue)	Cs,Sr (short-term heat mgmt.)	FPs (including lanthanides)	Pu+Np (for FR recycle fuel)	Am+Cm (heterogen-eous targets)
UREX+4	U (highly purified)	Tc, I (LLFPs, dose issue)	Cs,Sr (short-term heat mgmt.)	FPs (including lanthanides)	Pu+Np (for FR recycle fuel)	Am (heterogen-eous targets)	Cm (storage)

- UREX+1 and UREX+1a are designed for homogeneous recycling of all transuranics to fast-spectrum reactors.
- UREX+2, +3, and +4 are designed for heterogeneous recycling, possibly as an evolutionary step, to preclude the need for remote fabrication of fuel.

* SOURCE: Laidler, 2007.

Table UREX+ Variants and Their Associated Technologies and DOE-assessed Technological Maturity†

Process	*Fuel Type*	*Fabrication Technology*	*Technological Maturity*
UREX+1	(Interim storage only)	-	-
UREX+1a	FR mixed oxide	Remote, hot cell	Low
UREX+1a	FR metal	Remote, hot cell	Low
UREX+2	(Interim storage only)	-	-
UREX+3	LWR mixed oxide	Glovebox	High
UREX+3	FR mixed oxide or metal	Glovebox	High
UREX+3	Am/Cm transmutation target	Remote, hot cell	Low
UREX+4	LWR mixed oxide	Glovebox	High
UREX+4	FR mixed oxide or metal	Glovebox	High
UREX+4	Am transmutation target	Remote, possibly glovebox	Low
UREX+4	Interim storage of Cm	-	-

† SOURCE: Finck, 2006.

The joint committees' statement of task calls for a comparison of the PUREX reprocessing process, the UREX family of processes, and other spent fuel treatment processes being considered or developed in the United States and Russia. The joint committees found that insufficient information was available for realistic comparisons. First, as noted above, meaningful comparisons consider entire nuclear energy systems, rather than being based on a single component of those systems, such as fuel processing. Second, while PUREX is an established industrial process that has been used at a large scale for decades in several countries, the UREX family of processes is still at an early stage of development, and the features depend very much on the details of the process and the fuel to be processed.[4] PUREX itself is not a single process but a series of solvent extraction steps with several variants, somewhat different in each incarnation.

The development and selection of the technology options requires clear goals. PUREX was initially developed to separate high-purity plutonium for nuclear weapons. Variations on PUREX may try to improve the process with respect to other objectives, but the process inescapably bears some features of that original design goal that both cause proliferation concerns if the technology spreads and result in waste streams that have proven problematic. Alternative methods for processing are being designed to other goals, but those goals are not always clear or compatible. Key decisions about the specific process under consideration, whether PUREX, UREX, or some other process, strongly affect issues such as the radiation levels from the materials to be recovered for recycling and the characteristics of expected waste streams. In general, the UREX family of processes involves additional separation steps not included in PUREX, and is therefore likely to be somewhat more complex and expensive, and may increase the difficulty of material accountancy, though there may be potential for further optimization of whatever processes are eventually developed. Studies to date suggest that the material recovered for recycling in these processes would be more radioactive than the plutonium recovered in the PUREX process, but not radioactive enough to be a substantial barrier to theft and subsequent processing for use in a nuclear explosive.

Pyrochemical processes have been pursued in Russia and the United States, and elsewhere, with a particular emphasis on processing fast-reactor fuels. Russia's process is well along in development, and Russia has decided to use this process for processing spent fuel from the BN-800 fast reactor now under construction.

Russia has decided to use pyroprocessing combined with vibropacking the fuel to produce assemblies for the BN-800 fast reactor. These technologies complement each other well and produce fission materials with a sufficiently high level of radioactivity at each processing stage, in a mixture with minor actinides and certain fission products. Its high radioactivity drives the application of remotely controlled and fully automated fuel manufacturing processes in a closed fuel cycle, so that the fuel is very difficult and very costly to remove for other purposes.

There is far less experience with the Russian pyrochemical process than there is with PUREX, however, and estimates of costs for widespread deployment are still difficult to make. It appears that the wastes from the process can be made suitable for geologic disposal. The material recovered from the Russian process, sometimes called dirty fuel in Russia, includes

---

[4] "The characteristics, treatment, and final disposition requirements of several waste streams from spent fuel reprocessing is not completely known at this time. This is because (a) different separations and fuel fabrication options are still being evaluated, (b) waste stream generation from the proposed separations options is uncertain and unprecedented, and masses, volumes, and compositions remain uncertain… The UREX suite of separation technologies can result in many different waste streams." (DOE, 2008a, p. 24).

several actinides and some fission products, and Russian sources report that even a kilogram of the recovered material emits several Sv/hr, which is above the international standard for self-protection. Most of this radiation, however, comes from fission products with short half-lives. The radiation barrier that remained if the material were stored until the short-lived fission products decayed would still be too high for hands-on operations in normal commercial environments, but not too high for determined terrorists to attempt to use the material for weapons. (The same is true of commonly discussed variants of the UREX process.)[5]

The joint committees believe that additional fuel treatment processes, not currently being actively pursued in the United States or Russia, deserve additional exploration, including, for example, processes making use of supercritical CO_2 and fluoride volatilization. Decision makers still need to know whether these processes can overcome any of the most important cost, proliferation, safety, and security issues associated with the traditional PUREX process.

The joint committees' statement of task also calls for a comparison of the Russian BN-600 and BN-800 fast reactors to the types of fast reactors under consideration in the U.S. Global Nuclear Energy Partnership (GNEP) Program. This is, in a sense, an apples-to-oranges comparison, as these reactors are at very different stages of development and being pursued with very different purposes in mind. The Russian reactors are breeders, designed to produce more plutonium than they consume to address long-term concerns over limited uranium resources, while the proposed GNEP concepts are burners, designed to burn up stockpiles of plutonium and other actinides in the minimum number of cycles.[6] The BN-600 reactor has been operational for decades, and the BN-800 is under construction, while the proposed GNEP reactors are still paper concepts. While the BN-600 and BN-800 reactors have breeding ratios just over 1.0, some Russian designers envision future reactors with breeding ratios in the range of 1.6, which would be a major technical challenge; GNEP, by contrast, envisions burners with conversion ratios in the range of 0.25-0.5, also a major technical challenge. While the number of cycles required to achieve any given level of actinide destruction can be calculated for burners of any given conversion ratio, it makes no sense to compare the proposed GNEP reactors to the Russian designs in this respect, since the Russian designs are not intended for this purpose. As currently planned, the BN-800 will operate with oxide fuel, and the spent fuel will be pyroprocessed and new fuel produced with a vibropack process, demonstrating these approaches on an industrial scale. (More detail on these Russian fuel cycle plans is provided in Appendix C.)

With advanced computer modeling, it may be possible to design a fast reactor for which it can be demonstrated that the reactor would shut itself down automatically in response to any of the plausible transients in the system; if so, this would be a substantial safety advantage, and might make it possible to eliminate some of the redundant safety systems now used in light-water reactors, potentially reducing costs. This possibility and the cost impacts that might result from it, however, both remain to be demonstrated.

Here, too, the joint committees believe that continued funding for research and development on fast-reactor concepts and other reactor types not currently being actively

---

[5] PUREX was designed to separate out plutonium, which is a nuclear weapons material.

[6] The Russian fast reactors can be configured to burn (that is, have a conversion ratio of less than 1), but they have not been operated in that way or with minor actinide-bearing fuels and are not optimized for this configuration and mode of operation. Some fuel tested at the Russian BOR-60 test reactor with minor actinides (neptunium and americium) were "semi-industrial" rather than laboratory studies. Many difficulties arise, however, in building commercial-scale facilities even with semi-industrial-scale experience.

pursued in the United States and Russia would be desirable, including such concepts as lead-cooled systems, nonfertile fuels, thorium fuel cycles, and molten salt reactors.

Finding 8a

Both Russia and the United States are working on new technologies for processing spent fuel, intended to reduce the economic costs and proliferation risks of traditional reprocessing approaches and improve waste management. The technologies being proposed would still pose significant proliferation concerns if deployed in countries that did not previously have reprocessing capabilities. The new technologies under development will take significant time before being ready for demonstration at commercial scale.

Finding 8b

In most cases, reprocessing is not economic under current conditions. When the world's economically recoverable uranium resources diminish compared to demand or there is widespread deployment of fast reactors, then reprocessing may become economically attractive.

Recommendation 8

Developers of nuclear fuel cycle technologies should assess the technologies' proliferation risks and projected economic costs and benefits as critical elements of design.

As new technologies are developed, it will be important for developers to consider the proliferation hazards and work with the IAEA to develop appropriate safeguards.

Finding 9

Excess stocks of plutonium separated from spent fuel, beyond plutonium that would be needed for making MOX fuel for use in the near term, pose security risks.

Recommendation 9

States should end the accumulation of stockpiles of plutonium separated from spent fuel as soon as practicable, and begin to reduce existing stocks. Spent fuel should only be reprocessed when its constituents are needed for fuel, or when reprocessing is necessary for safety reasons.

WHY "ACCEPTABLE LEVEL OF DESTRUCTION OF ACTINIDES" IS NOT WELL DEFINED TECHNICALLY

Actinide destruction, more properly *actinide fissioning* or more commonly *actinide burning*, has been stated as one of the main objectives of the advanced technologies for nuclear energy in the United States, and has been considered as a central objective of programs in Japan and Europe. As articulated in GNEP, actinide burning is meant to support three main goals: extracting more energy from the earth's uranium resources, reducing the quantity and hazard of radioactive waste in a deep geologic repository, and reducing the potential for fuel cycle material to be used to make nuclear weapons. In the joint committees' view, each of these is a worthy goal. They cannot, however, simply be addressed by pursuing actinide burning.

The actinides are not a single species. The specific goals for the various species differ depending on the larger fuel cycle system into which actinide burning is being deployed. While all of the transuranium actinide nuclides can undergo fission, some are more useful for reactor systems than others, and some reactors are better matched with particular nuclides over others (more on this below). Similarly, in a geologic repository, some nuclides are greater contributors to risk than others, and which ones are the main contributors to risk depends on the repository system design and environment. The quantity of waste that can be loaded into a repository depends in part on the heat output of the waste to be emplaced, and also on the characteristics of the repository system. And finally, the technical objectives to serve nonproliferation and safeguards depend on the kinds of scenarios that cause concern. Radiation barriers and the presence of other actinides with plutonium in a material stream could present significant obstacles to terrorist groups, but are unlikely themselves to be major obstacles for a nation seeking nuclear weapons.

Without a clearly articulated technical objective, there is no credible technical basis for answering the question, What is an acceptable level of burn-up? Burn-up is expressed as either the fraction of the initial heavy metal that has fissioned or as the energy released per ton of initial heavy metal in the fuel (for example, 5 percent or about 55,000 MWd/MTHM). This single number, however, is unlikely to provide the information needed about how the actinides have been burned for a given goal.

Actinides, and even isotopes of the same actinide element, do not burn uniformly. The nuclides have different reaction rates[7] for a given neutron spectrum, so adjacent nuclei of plutonium-239 and plutonium-240 will be destroyed at different rates in the same neutron flux. The reaction rates are functions of

- the cross sections (essentially the reaction probabilities), which are fixed for a given neutron flux spectrum
- the concentration of the nuclides
- the neutron flux spectrum and magnitude (energy distribution and strength)

But as the reactions change the concentrations of different nuclides, this affects the neutron spectrum, which changes the relevant cross sections, and all of the reaction rates change. The neutron flux and energy spectrum, then, cannot be seen as fixed for a given reactor because it evolves with the changing composition of the fuel, and varies spatially across the reactor core. Finally, the neutron energy spectrum is different for different reactors depending on the coolant (for example, sodium or lead), the fuel type (for example, metal or oxide, and precisely which metals), the cladding, the operating mode, and the configuration (surrounded by a reflector, a fissile blanket, or a fertile blanket). While all reactors have a range of possible neutron spectra, that range tends to be larger for fast reactors.

A reactor using fuel initially loaded with 15 percent plutonium that reaches high fuel burn-up, for example, 135,000 MWd/MTHM (around 15 percent burn-up), may still have substantial amounts of plutonium and other actinides in the spent fuel. This is because not all of the fissions occur in the plutonium. In uranium-plutonium dioxide (U-PuO_2) fuel, most of the

[7] The term burn is a nontechnical reference to fissioning a nucleus, but the nuclei also undergo other reactions that do not burn that nucleus. These reactions compete with each other for each impinging neutron, and while some actinide nuclei will split, others will be transformed into other heavy nuclides.

fissions occur in the plutonium, but some of the fissions occur in the uranium, and neutrons absorbed in the uranium create more plutonium and other heavy nuclides that may fission or remain in the spent fuel.[8] Such a reactor can be designed with a blanket of natural uranium or depleted uranium that produces plutonium, and overall the system may have more plutonium (more higher actinides) after a cycle than it had at the beginning of the cycle. One scheme for effecting improved burnout of particular species, such as neptunium and americium, is to separate them when processing spent fuel and load them into burning targets distinct from the reactor fuel. These targets could have compositions tailored to the chemistry and neutronic characteristics of the actinide atoms they hold. The parts of the core may then have modified spectra that are better for burning the targets, and the targets could reside in the core for longer or shorter times than a fuel assembly does. The range of required irradiation times or fluxes might dictate (economically) different target hardware, such as cladding.

For systems that are designed to multicycle fuel, almost certainly different isotopes will require a different number of cycles to achieve desired reductions in actinides inventory. For such systems, analysts must also examine the efficiency and effectiveness of the processing facility. The product stream and the waste stream both matter for the resource utilization, waste hazard and footprint, and nonproliferation goals.

At a more basic level, many of the critical reaction cross sections, and their variation with spectrum, are currently based on theory and analysis rather than measurements. Experimental measurements would seem to be essential before proceeding to actually plan the program. This issue could in fact have a major impact on the preferred type of reactor to build. The location in the reactor where the targets are placed will have a major influence on the results.

In developing reactors for recycling of actinides, the United States and Russia have focused on sodium-cooled reactors. The burner reactor currently proposed in GNEP has not yet reached conceptual design (the decision whether to use metal or oxide fuel, which is essential to the design of the reactor core, has not been made). The Russian BN-600 has operated for 25 years on HEU fuel with some tests using MOX fuel. The BN-800 is under construction and is anticipated to be commissioned in 2012, and the government recently decided that the reactor will operate with vibropacked MOX fuel. It will start with recycled weapons plutonium. There is not now a plan for it to burn any of the higher actinides.

B3. What impact could new technologies have on these proposals?

The present commercial nuclear fuel cycle includes the mining and extraction of uranium, the purification of uranium ore, the conversion to uranium hexafluoride, uranium enrichment, fuel fabrication (including the conversion of uranium hexafluoride to uranium dioxide). Irradiation in a nuclear reactor is then followed by storage and either reprocessing or disposal of the irradiated fuel. France now accomplishes what GNEP envisions for the United States, to recycle some of the by-product plutonium in light-water reactors (LWRs) as mixed-oxide fuel (uranium dioxide–plutonium dioxide [UO_2-PuO_2]). It is postulated that in the future

---

[8] An alternative is to use a different fuel material, such as thorium-zirconium-plutonium. Irradiation of thorium produces fissile uranium-233, but does not produce appreciable amounts of the higher actinides while it burns out the plutonium and its activation products. This could reduce the weapons-usability of the material dramatically and could reduce the quantities of radionuclides that dominate risk from some repositories (for example, neptunium-237). Its benefits with respect to other goals, resource utilization, and repository heat load, are less clear. Please see below for further discussion of this point.

fast-neutron reactors will be used both to produce fuel (plutonium from uranium-238 or uranium-233 from thorium) and to burn up (destroy by irradiation) minor actinides and some fission products. The fuel cycles of these fast reactors will be quite different from those of LWRs. There would be no need for uranium enrichment and much less need for uranium ore, but there would be a need to process breeder reactor blankets, and possibly to process burn-up targets, and irradiated fuel (possibly three or more quite different processes). All aspects of both the current fuel cycle and the future postulated fuel cycles are subject to significant changes due to new technologies.

New, improved technologies could increase the attractiveness of nuclear power and support its widespread expansion. Although no technologies themselves solve the fundamental problems of internationalized fuel cycles, new technologies could improve nuclear fuel cycles in utilization of fuel resources, reducing quantities and hazards of radioactive waste, broadening the options for a high-level waste repository, making nuclear reactors more economically competitive in more situations, and reducing the proliferation hazards associated with fuel cycles. The critical questions about new technologies are how much of an improvement do they make, what new risks might they pose, over what time frames could they be realized, and do these improvements make a difference in the overall desirability of future fuel cycles? The joint committees can only begin to offer answers to these questions and make some general observations.

First among the observations is that these topics are not areas of technology that will advance without directed research specifically focused on the nuclear fuel cycle; advances in other areas of science and engineering will help, but are not sufficiently linked to nuclear fuel cycles to solve the technical challenges described here. Research is needed in the areas of processing of irradiated nuclear fuel and nuclear fuel design, as well as in improved approaches to disposal of wastes or spent fuel, and reduced-cost recovery of uranium from low-grade sources.

Nuclear science and technology has reaped great benefits from improvements in computing for simulations and control systems and improvements in techniques for fabrication and processing materials (for example, improving the corrosion resistance or hardness of metals, and high-accuracy manufacturing or machining to very small tolerance). But many of the science and technology needs within the nuclear energy sector, especially those affecting the improved design and fabrication of nuclear fuel, reactor design and operation, and irradiated fuel processing, are unique to this sector. The relevant nuclear reaction cross-section data are unlikely to be gathered for other industrial sectors. Satellite manufacturers and space agencies are interested in "radiation resistant" materials, but making materials that are durable in a reactor environment is a challenge that is unique to nuclear reactor engineering. Irradiated nuclear fuel comprises a diverse set of chemical constituents. Carrying out separations on these constituents, some of which are processed in no other context, in the high-radiation environment they generate poses a unique challenge. These challenges belong squarely to the nuclear energy sector, and a committed effort to research and development in nuclear science and technology is needed to make progress on them.

Some nations—notably France, Japan, Russia, and South Korea—have made progress in development of nuclear reactors, nuclear fuels, nuclear fuel processing, and their nuclear energy enterprise. The UK experience with reprocessing has not been favorable.[9] The governments of

[9] Martin Forwood reports that "After a projected slow ramp-up period, THORP was to achieve 900 tons/yr in the sixth year of operation (1999) with fuel burn-up ranging from 3.1 to 24.0 GWd/MTU for AGR fuel, 7.4 to 28.8 for

these nations have been committed to making progress, in some cases at great cost. Russia has made important progress on several fronts, creating institutional structures and arrangements in parallel with commitments to expand its use of current-generation and next-generation nuclear power plants, and focused research and development to address several technological challenges. Many other nations have shown less consistent support and little commitment to nuclear energy. The United States increased its commitment in the first half of this decade. The domestic component of the U.S. Global Nuclear Energy Partnership would require substantial funds to achieve its ambitious goals. A recent National Research Council report (NRC, 2007) criticized this program largely because to achieve almost any combination of its stated goals, the program would have to rely on new technologies, but the program was framed to move rapidly toward construction of facilities using near-term technologies. This brings us back to the critical questions about new technologies: How much of an improvement do they make? Over what time frames could they be realized? Do these improvements make a difference in the overall desirability of future fuel cycles?

The following discussion describes several areas of technology in which improvements could have a substantial impact on the options available for international fuel cycles: Better fast reactors; small, self-contained reactors; new technologies for enrichment; high burn-up fuels (one-pass reactors and multirecycled transuranic fuels); thorium fuel cycles; dry methods for fuel separations; and economic new sources of uranium.

IMPROVED FAST REACTORS

It may be possible to show that some fast-reactor designs would not require some of the safety systems required for LWRs. Some reactor experts have long suspected that fast reactors, through choices of fuel, configuration, and coolant, can be designed to be passively safe[10] and store a relatively small amount of energy in their primary coolant systems. This could not be satisfactorily demonstrated in the past,[11] but with a combination of new experiments and the computing power now coming available, more definitive answers to these questions may become available. If indeed reactors can meet these hopes, then in addition to the inherent benefits, the reduction of required safety systems might make these fast reactors cost-competitive with LWRs.

BWR fuel, and 16.9 to 40.0 for PWR fuel—with higher burn-up fuels reserved for later years of the base-load..... [T]hroughput has neither been reliable nor to specification—with just over 5,000 tons completed during the first ten years of operation. Contributing to schedule slippage have been a range of equipment failures and accidents including acid spills, pipe leaks and blockages and problems with the plant's sole high-level waste evaporator. By the end of the official base-load period, with plant closure scheduled for 2010/11 'with all contracts completed,' THORP was running some two years late. No new orders were secured and none are currently in the pipeline." (Forwood, 2008, p. 20). "The UK reprocessing program has produced an accumulated separated plutonium stock of over 100 tons as of the end of 2006. It is considered to be an asset of 'zero value' and it is as yet undecided whether the UK will treat it as a waste product or a future energy asset. The stockpile will increase to 133 tons if current reprocessing contracts are completed. Some 100 tons will be from UK-origin spent fuel and 33 tons from foreign fuel. The plutonium from foreign spent fuel is to be returned to overseas customers as MOX fuel," (Forwood, 2008, p. 35).

[10] Here *passively safe* means that the reactor has negative reactivity coefficients in cases of voids and temperature rises in the fuel and in the primary coolant, and that the primary coolant can remove sufficient heat from the core in believable accident scenarios.

[11] Experiments on EBR-II demonstrated performance only for metal fuels.

SMALL, SELF-CONTAINED, DEPLOYABLE REACTORS

Reactor designers in several countries have been developing designs for small- and medium-sized reactors with improved nonproliferation features for deployment in developing economies and specialized applications (IAEA, 2005a). An interesting subset of these is factory built and fueled reactors. All commercial, light-water power reactors worldwide have historically been built at the site of operation and are refueled by opening the reactor, removing fuel that is spent and placing it in temporary storage, and loading the reactor core with fresh fuel. For a typical light-water reactor core, the reactor must be shut down every 18 or 24 months to change out one-third of the fuel. Some other power reactors (RBMKs, the CANDUs, and some others) refuel online without opening the reactor. But different designs for small reactors enable the reactor vendor to construct the reactor core and primary coolant system at a factory and then deliver this self-contained unit to the site of reactor operation, which may be on land or on an offshore floating power station. The size of the reactor and its weight would allow vendors to deliver modules by river barge, rail, or airplane. With a long-life core (ranging from 7 to 25 years, based on design), the unit can be provided fully fueled and need never be refueled at the site of operation. There are safeguard advantages to a reactor that is sealed from the day it is made at the factory until the day it returns to the factory, at the end of its useful life. As noted elsewhere in this report, under a fuel-leasing agreement, the irradiated fuel would not remain in the lessee's possession, and so fissile material could not be separated from the fuel without breaching the agreement. Under a reactor-leasing agreement, the fuel would never leave the reactor core, which makes breaching the agreement even more readily detected. In part for these reasons, the GNEP goals included a call for small reactors, although the U.S. government has not sponsored much work in this area.

The Russian Federation is much further along in developing compact, factory built reactors than any other country, having nearly completed construction of a floating demonstration reactor in Severpodvinsk in the Arkhangelsk Region based on one of its several detailed designs (KLT-40C, which has a 3-year refueling interval; see Table 3-1). The economic viability and attractiveness of these reactors will become clearer after one or more customers have operated them for more than one fueling cycle. The cost of electricity from such reactors right now is expected to be substantially higher than electricity from fossil fuel plants and more conventional nuclear power plants, so Russia is marketing them for specialized situations and applications. For example, mining and refining operations in remote locations are potential niches, as delivery of fossil fuels in Siberia is costly, and use of natural gas for recovery of oil from tar sands in Canada is inefficient. Remote populations could also be markets for district heat, desalination, or electricity. Countries that want nuclear power on a small scale might find attractive these simplified reactors with passive safety systems that could function in environments where there is not a highly developed nuclear infrastructure. Such reactors might become more attractive if production economies of scale bring the prices down.

Japan's Toshiba is also marketing a small sodium-cooled reactor with a long-life core. Toshiba 4S (Super Safe, Small, and Simple) nuclear power system is a 10-MWe reactor designed to operate for 30 years without refueling. The fuel is 24 percent uranium and 10 percent plutonium in a zirconium matrix. Like the Russian design, the 4S (also called the nuclear battery) could be transported in modules by barge. A remote community in Alaska in the United States is considering purchasing one of these power systems.[12]

TABLE 3-1 Small Reactors

Small NPP	Power Capacity Electricity, MW	Cogeneration Electricity, MW	Cogeneration Heat, Gcal/h	Refueling interval, years	Fuel enrichment, %
ABV-6	2x8.5	2x6	2x12	12	19.5
SVBR-10	2x12	2x6	2x25	12	18.7
Uniterm	2x6.6	2x2.5	2x17.2	25	19.5
KLT-40C	2x38.5	2x19.5	2x73	3	17.4
Ruta	-	-	60.2	3	3
VVER-300	300	220	450	2	3.3
VBER-300	2x340	2x215	2x460	1.5	19.5
VK-300	-	250	400	2	4
SVBR-100	4x101.5	4x95	4x130	8	16.5
4S	1x10			30	Pu-10%

SOURCE: Adapted from IAEA 2005a.

HIGH BURN-UP FUELS

Advanced fuel technologies could have an impact on the options available for nuclear fuel cycles in that they are essential to the technical feasibility of several of the options. As is described elsewhere in this report, nuclear fuel cycles need to be considered as systems and evaluated against criteria or goals. Different nuclear fuels and fuel cycles can accomplish different goals in different ways. A fuel cycle may achieve high energy utilization by achieving high burn-up in the fuel in a once-through fuel cycle (such as plutonium-thorium-zirconium fuel for sodium-cooled reactors and silicon-carbide–coated [Si-C–coated] fuel for gas reactors) or by multirecycling fuel (transuranic fuel from breeding and burning fuel cycles). The fuel matrix may be metal, oxide, nitride, or carbide, or even some dispersion combination with other materials. These examples are given as illustrations; among experts there are different opinions about each fuel's state of development and which options best fulfill specified goals. Reactor designers must overcome significant materials challenges with the assistance of materials

[12] See Overview of Galena's Proposed Approach to Licensing a 4S Nuclear Reactor Based Power Generation Facility, 2008.

scientists before any of these fuel technologies can be demonstrated to fulfill their promise and be deployed.

By *high burn-up fuel*, the joint committees mean fuel that achieves more than 200 MWd/kg, or at least four times the level of current light-water reactor fuel. High burn-up fuels must have two characteristics: They must be able to endure the physicochemical and radiation conditions over their long irradiation lifetimes, and they must contain or make enough fissionable material to sustain energy production over the fuel lifetime.

Each of the fuels mentioned above has different advantages: A long-life plutonium-thorium-zirconium fuel would burn out nearly all of its fissile plutonium isotopes and sustain itself on uranium-233 bred from the thorium-232 in the fresh fuel. The proliferation aspects of uranium-233 fuel are described later in this report. This fuel is highly refractory, and it is therefore difficult to separate its constituents. Silicon-carbide–coated fuel spheres are likewise difficult to process for separations, and they have attractive features for reactor safety (especially the high coefficient of thermal expansion, which affects reactivity, and very high melting temperature) and are durable even under extensive irradiation. Both fuels can accommodate a range of actinide compositions and can be designed to accommodate fission products that accumulate during irradiation, and their robustness enables them to survive the radiation damage.

Multirecycling that burns the actinides requires transuranic fuel; that is, a fuel with high transuranic content that maintains stability but can also be processed for separations. Setting aside economic viability of these systems, design and fabrication of such fuel has been identified as the greatest technical challenge for fuel cycles considered under the advanced nuclear energy development program proposed in the United States in recent years. A system that retains the higher actinides within the fuel materials to reduce the direct usability of the materials streams in weapons and to reduce the actinide content of the waste streams faces the challenge of creating fuels that have never been fabricated and run before. For metal fuels, a major challenge is retaining americium and curium in the fuel during fabrication because those elements are volatile in the temperature range in which the tested fuel fabrication techniques operate. There is some experience with the EBR-II from the 1960s and 1970s, but this provides only a starting point and some lessons. For oxide fuels, the fabrication problems are somewhat easier but still have not been demonstrated. A heterogeneous reactor core containing both fuel rods and target rods (containing the americium and curium) would simplify meeting the actinide burning goal, but compromises on the goal of maintaining the actinides together in the materials streams.

BOX 3.4
SAMPLE LESSONS FROM THE EXPERIMENTAL BREEDER REACTOR II

By Milton Levenson

Perhaps the most important lesson from the Experimental Breeder Reactor II (EBR-II) was that to be successful the entire system had to be considered in design and operation. Originally EBR-II was three independent projects—the reactor, the fuel, and the fuel cycle—each being pursued in a different division at Argonne National Laboratory. After the completion of the Title I design, Dr. Walter Zinn, director of Argonne, instituted a single design review of all three projects at the end of which he combined the three into a single project under a single project manager. During the combined review many changes were made in all aspects so as to optimize the total project. Some examples include the following.

Before the reactor was built, an analysis was done to determine the fuel composition after infinite recycling, considering the residues that would be left in the fuel by pyroprocessing. This composition, called fissium, was then the original fuel composition using stable isotopes in the original fuel. By this means there was very little change in the fuel's chemical composition with recycling, and the process always saw the same material as feed.

The fuel was considered as a three-component system with three services: (1) to power the reactor, (2) to be remotely disassembleable, and (3) to be remotely fabricable. The components were the fuel matrix, the clad material, and the geometry. The final fuel was different from any original design, but one that satisfied all needs. The very high burn-ups (200,000 MWd/MTHM) were achieved by adjusting all three. Fuel to clad gaps were adjusted so that the fuel could swell only enough to allow fission gas bubbles to connect and so vent to the cladding. This stopped the swelling pressure that otherwise would rupture the cladding. Research continued on both cladding and fuel, but not independently.

To address the question as to whether a "vented" fuel element might be used in the future—or whether a cracked cladding might be considered dangerous—one subassembly was fabricated without any cladding and run in the reactor at full power with no significant adverse affects. Some iodine migrated into the sodium as sodium iodide and was removed in the cold trap. Some xenon and krypton migrated into the cover gas closed system. However, this success is relevant only to a metal fuel alloy that is thermodynamically stable when in contact with sodium, which illustrates the point about a systems-design approach.

In addition to the fuel matrix itself, other fuel materials must be able to perform reliably throughout the fuel's residence within the reactor core. Most fuels[13] have a metal casing, called cladding, that separates the fuel matrix from the primary coolant. Cladding performance has been the limiting factor for burn-up of fuels in the past. Minor defects in manufacturing of cladding can lead to cladding breaches (failed fuel), which releases radioactive material into the coolant and allows the coolant to interact directly with the fuel, as happened commonly with boiling water reactors until the 1990s. Even as manufacturing quality improves, however, reactor designers and operators must grapple with more fundamental limitations of the cladding materials, as the accumulated radiation damage for high burn-up fuel exceeds the cladding's ability to self-anneal and maintain its integrity. Mechanical damage, too, can be a factor, as pressure and flow-induced vibrations strain cladding and other fuel materials.

[13] The silicon-carbide spheres mentioned above are an exception, although some of these fuels are encased in larger graphite shells. Liquid fuels such as molten salts are another exception.

For the postulated fast-reactor cycle, new technology is required to bring it into being. Such cycles are much more interactive than those of the LWRs: The fuel, the cladding, the blanket, the coolant, the fuel processing, and the blanket processing must be treated as a single system if the result is to be an effective solution (see Box 3.4 for some lessons learned from EBR-II). Significant pieces of a fast-reactor fuel cycle do exist, and the proof-of-principle as a power source has been established, but there does not exist a fuel-reactor-process integrated system that can be demonstrated to be either economic or operationally successful as long as there is not an agreed set of metrics and criteria that defines success.

THORIUM FUEL CYCLES

The use of thorium in reactors has been studied for several decades. The AVR 15 MWe experimental pebblebed reactor operated at Julich, Germany, from 1967 to 1988. Based on the AVR, the 300 MWe thorium high-temperature reactor (THTR) in Germany operated from 1983 until 1989. In the United States, the Fort St. Vrain 330 MWe high-temperature gas-cooled reactor (HTGR) operated from 1978 to 1989 with thorium/HEU fuel. It never operated well, but the problems were not associated with the use of thorium in the fuel. Also in the United States, the Shippingport LWR operated as a breeder reactor from 1977 to 1982 using the Radkowsky seed-and-blanket design. In India, Kakrapar 1 and 2 use some thorium fuel (WNA, 2008).

World resources of thorium are four to five times greater than those of uranium. Introduction of thorium fuel cycles would tap those resources for power generation and could reduce the waste disposal and proliferation hazards of nuclear power engineering, depending on how such cycles were implemented. Neither uranium-233 nor plutonium is found in significant quantities in nature, and so they must be produced in a reactor to acquire enough material to fuel a reactor.

One approach that has been proposed in recent years is to use uranium-thorium seed-and-blanket fuels in existing LWRs. This would result in lower total quantities of spent fuel per unit of electricity generated and lower inventories of plutonium isotopes, making the spent fuel a less attractive source of plutonium for weapons (see, e.g., Galperin and Todosow, 2001). Another approach that has been discussed extensively is to use thorium-fueled molten salt reactors, with continuous partial removal of fission products, so that no weapons-usable uranium-233 is separated from the liquid fuel (see, e.g., Gat et al., 1993a). Moreover, due to the absence of uranium-238, accumulation of minor actinides in the thorium fuel cycle is considerably slower than that in the uranium fuel cycle.

Thorium-uranium-233 fuel cycles generally build up small concentrations of uranium-232, whose decay products, bismuth-212 and thallium-208, emit hard gamma rays. This would mean that those who worked for many hours with typical uranium-233 would receive doses beyond worker health and safety limits, so a commercial operation, even at the early stages of introduction, would require the use of heavy biological radiation shielding and automated remote-controlled equipment. But the uranium-232 concentrations are typically insufficient to prevent advanced states from making weapons from these materials using frequent worker rotation insufficient to prevent terrorists from making crude bombs from uranium-233. Uranium-233 has a much lower critical mass than uranium-235, and unlike plutonium, it has low-enough neutron generation that it can be used in simple gun-type bombs. It is therefore a dangerous potential nuclear weapons material requiring stringent security and accounting measures.

Advances in aqueous and nonaqueous technologies could be introduced in future thorium fuel cycles. As noted earlier, there are no significant uranium-233 resources in nature, so for a thorium fuel cycle to be introduced, a plant must breed uranium-233, initially relying on some other fissile fuel to power the reactors. The accumulated uranium-233 can then be used to fuel industrial fast and thermal reactors.

In the initial stage of thorium fuel cycle development, Russia envisions reprocessing the following fuel types: thorium metal or thorium dioxide (ThO_2) of the BN-800 reactor blanket; PuO_2-ThO_2 fuel of the VVER-1000 reactor; $^{233}UO_2$-$^{235}UO_2$ fuel (hereinafter UO_2 fuel) of the VVER-1000 fuel; UO_2-ThO_2 fuel of the VVER-1000 reactor.

The purpose of reprocessing a fast-breeder-reactor-irradiated thorium blanket is to extract uranium-233 and return the remaining thorium back to the reactor for further irradiation. Other fertile fuels rely on fissile plutonium or uranium as the initial source of fission reactions in the fuel. PuO_2-ThO_2 fuel for thermal reactors can extend the use of the fuel with good nuclear-physical properties, burning out the plutonium without uranium-233 extraction. Fuel composed of $^{233}UO_2$-$^{235}UO_2$ can be reprocessed to purify uranium from minor actinides and fission products and recover uranium reactivity by adjusting its isotopics. Finally, UO_2-ThO_2 fuel can be reprocessed to extract uranium from minor actinides and fission products and adjust fuel reactivity according to the thermal reactor requirements.

For the second stage of Russia's thorium fuel cycle development, fast and thermal reactors with UO_2-ThO_2 fuel are being explored. The BN core is expected to operate using fuel containing minor actinides and long-lived fission products. In this case, the purpose of fuel reprocessing is its purification and adjustment of the composition to conform to the requirements of a uranium-233 fueled reactor. The uranium-233 accumulates in the fast breeder reactor alongside the metallic thorium blanket. The purpose of the BN metal blanket reprocessing remains the same, that is, extraction of uranium-233 and return of thorium into the reactor for further irradiation in the blanket. However, in a closed thorium fuel cycle, the purpose of thorium-blanket reprocessing is only to increase the uranium-233 fraction to the requirements of the reactor for which this fuel is produced.

"Dry" technologies of a thorium fuel cycle can be based on the following processes:

- hydrogenation of metallic fuel
- chlorination of metallic and oxide fuel
- sublimation and vacuum distillation of thorium and uranium tetrachlorides
- electrolysis of molten salts
- concentration of minor actinides and fission products
- production of fuel compositions, fuel elements, and fuel assemblies

These options are summarized in Table 3-2, which presents the types of reactors to be implemented during both stages of thorium fuel cycle development in Russia, and the purpose of reprocessing the blanket and core fuel.

TABLE 3-2 Reactor Types, Fuel Types, and Purposes of Fuel Reprocessing in the Thorium Fuel Cycle

Stages of Thorium Fuel Cycle	Reactor type, fuel accommodation		Fuel	Purpose of reprocessing blanket and core fuel
Stage 1 Accumulation of U-233	BN-800	Blanket	Metallic Th	Extraction of U-233
			ThO_2	
		Core	MOX	Recovery of NPhP
	VVER-1000	Core	PuO_2-ThO_2	Extraction of U-233, Recovery of NPhP
			$^{233}UO_2$-$^{235}UO_2$	Recovery of NPhP
			UO_2-ThO_2	Recovery of NPhP
Stage 2 Closed Thorium Cycle	BN-800	Blanket	Metallic Th	Extraction of U-233
				Adjustment of U-233 content
		Core	UO_2-ThO_2 + MA + LLFP	Recovery of NPhP, Introduction of MA and LLFP
	VVER-1000	Core	UO_2-ThO_2	Recovery of NPhP

LLFP – long-lived fission products

NPhP – nuclear-physical properties such as the physical integrity and composition of the fuel matrix.

SOURCE: Provided by joint committees.

DRY METHODS FOR FUEL SEPARATIONS

The Russian nuclear effort in dry methods for separation of nuclear fuel constituents is divided into two main categories: (1) pyroelectrochemical, which are the most compact, but provide only partial separation and purification; and (2) halogenide distillation, which can achieve high levels of purification of uranium (mainly) and plutonium from fission products. An integrated technology—a combined reactor and fuel-processing unit, such as a molten salt reactor—has the advantage of easy fuel preparation and recycling, because the fluid nature of the fuel provides extra flexibility and a simpler back-end fuel cycle. The molten salt reactor concept appears to have substantial promise not only as a transmuter of transuranics, but also as an advanced TRU-free system operating with the uranium-thorium cycle.

Pyroelectrochemical Processes

Basic research on molten salt systems has enabled Russian facilities to develop processes for production of granulated uranium and plutonium oxides and mixed uranium and plutonium oxides. Pyrochemical technology is able to carry out all of the deposit production operations in one apparatus—a chlorinator-electrolyzer—which simplifies the process. Russian pyrochemical reprocessing consists of three main stages:

1. dissolution of initial products or spent nuclear fuel in molten salts
2. precipitation of plutonium dioxide or deposition of electrolytic uranium and plutonium dioxides from the melt
3. processing of the material deposited on the cathode or precipitated at the bottom of the melt for granulated fuel production

The process can recover the cathode deposits without changing their chemical composition or redistributing the plutonium. Three alternatives were considered and are now under development for reprocessing irradiated nuclear fuel at the Research Institute of Atomic Reactors (RIAR):

1. reprocessing uranium fuel with the production of uranium dioxide for recycling
2. reprocessing MOX fuel for only plutonium recycling as the most valuable component
3. reprocessing MOX fuel with production of MOX fuel

All products are reprocessed with the goal of having a complete recycle of plutonium, neptunium, americium, and curium. Vibropacking technology is applied to the manufacture of fuel pins.

VIBROPACKING PROCEDURE

RIAR has used vibropacking technology for about 20 years to fabricate granulated fuel in glove boxes or hot cells. The main advantages of the vibropacking technology and fuel rods with vibropacked fuel are as follows:

- The production process is simple and reliable because it has a relatively small number of subprocesses and control operations, which facilitates automation and remote control.
- The granular form of the fuel feedstock enables vibropacking technology to use both homogeneous compositions and mechanical mixtures for heterogeneous compositions.
- The thermal-mechanical stress on the cladding is lower for vibropacked fuel than for pellet-stacked fuel.
- Vibropacked fuel tolerates relaxed requirements for the inner diameter of fuel rod cladding.

Vibropacked fuel is made by agitating a mechanical mixture of (U, Pu)O_2 granulate and uranium powder, which binds up excess oxygen and some other gases (that is, operates as a getter) and is added to the fuel mixture in proportion during agitation. The getter resolves problems arising from fuel-cladding chemical interactions. The process allows fabricators to control the distribution of plutonium and density along the fuel column length, with the getter distributed uniformly throughout.

CLOSING THOUGHTS ON NEW TECHNOLOGIES

If we are to achieve anything with technology, what is needed is a set of specific objectives that can be used to guide the research and development programs.

Finding 10

Many of the technologies for improved nuclear fuel cycles are not areas that will advance without directed research specifically focused on the nuclear fuel cycle; advances in other areas of science and engineering will help, but are not sufficiently linked to nuclear fuel cycles to solve the technical challenges described here by themselves. Research is needed in the areas of processing of irradiated nuclear fuel and nuclear fuel design (beyond the incremental improvements in uranium oxide fuel for light water reactors), as well as in improved approaches to disposal of wastes or spent fuel, and reduced-cost recovery of uranium from low-grade sources. Additional research and development is also needed to develop advanced safeguards and security technologies that can provide increased capabilities to detect covert nuclear facilities; highly accurate near-real-time monitoring of material flows in bulk processing plants with reduced intrusiveness, increasing confidence that any diversion would be detected; low-cost real-time monitoring that would set off an immediate alarm if stored nuclear material were tampered with or removed; effective protection against sophisticated outsider and insider theft and sabotage threats at reduced cost; and design of facilities to simplify and increase the effectiveness of safeguards.

Recommendation 10

The U.S., Russian, and other governments should take the lead in a cooperative international effort to make additional research and development investment in advanced safeguards and security technologies. A focused effort should be made to make the results of this research and development available to the international community to ensure that new facilities are more secure and readily safeguarded. The international community also should adopt the philosophy of designing high levels of security and safeguards into new nuclear systems and facilities from the outset, including both the inherent technical characteristics of the process and the institutional measures to be taken.

Finding 11

It is not possible today to construct an entire, operational international fuel cycle program.[14] Such a program will have to be built incrementally. However, elements of that program currently exist and the groundwork for other elements has been laid.

Recommendation 11

For new technologies, the U.S., Russian, and other governments should

[14] This would be run internationally and include all elements of the fuel cycle.

- **continue to invest in research and development on advanced approaches to once-through and closed fuel cycles that offer the potential to improve proliferation resistance, safety, security, economics, resource utilization, and/or waste management.**
- **utilize a systems approach to developing and assessing these technologies, with clear objectives and technically justifiable criteria for decision making. Use systems analysis to identify potentially promising approaches before proceeding to build pilot or larger facilities.**
- **take all relevant proliferation risks into account when assessing proliferation resistance, including how the availability of the materials, facilities, and expertise associated with a particular fuel cycle approach would affect the time, cost, uncertainty, and detectability of a nuclear weapons program.**

The implementation of those elements that are feasible, for example, assurance of fuel supply, should not be delayed while other options are being refined or explored both institutionally and technically.

Secondary Issues:

B4. Compare the fuel to be produced from the processes examined in (B1) for use in appropriate reactors (light-water reactors, high-temperature gas-cooled reactors, and fast reactors). What are the advantages and disadvantages of each type of fuel?

B5. Compare the repository requirements for the waste produced by the processes proposed in the GNEP concept with that from a system based on PUREX and one based on Russian plans.

Handling the fuel after use in a reactor is difficult. Only Finland has an approved process to build a repository for spent fuel and Sweden may be close to having a site acceptable to a local community.[15] The three largest users of nuclear power, France, Japan, and the United States, do not have operating sites and only the United States has selected a site for a repository. Several billions of dollars have been spent in the United States, and on June 3, 2008, the U.S. Department of Energy submitted to the U.S. Nuclear Regulatory Commission an application for a license to construct a high-level radioactive waste repository at Yucca Mountain. The final standard for evaluating the license application has not yet been issued, and the regulator's review is still pending.

Although it is often argued that a closed fuel cycle reduces the volume of waste from nuclear energy, the amount of radioactive material requiring long-term storage depends upon the processes, the country's regulatory requirements, and even the definitions of waste.[16] Pool storage for 5 years followed by dry cask storage has been approved by the U.S. Nuclear Regulatory Commission as being safe storage for many decades. Nevertheless, some countries, such as France and Japan, are pursuing the option of reprocessing, which they believe offers

[15] Both Russia and, on a smaller scale, the United States have injected liquid radioactive waste underground as a means of disposal, but both countries now regard this practice as undesirable for future disposal.

[16] For an explanation and argument that the closed cycle produces more waste, see Schneider and Marignac, 2008.

waste management and resource extension advantages. Separating direct-use material by reprocessing significantly raises the proliferation risk from a nuclear program, but various forms of separation and recycling are nonetheless an important feature of some proposed fuel assurance programs. Countries embarking on nuclear energy programs should examine the approaches to management and disposal of radioactive wastes that they will pursue.

None of the fuel assurance programs discuss the possibility of taking the spent fuel to encourage a new country not to build a reprocessing plant. Unfortunately, except for the Russian program, there is little likelihood that any of the other programs will be able to offer to take the spent fuel. This gradually may become a difficulty in maintaining credibility of the programs.

COMPARISON OF PROCESSES FOR SEPARATION OF FISSILE AND OTHER MATERIALS FROM SPENT OR IRRADIATED NUCLEAR FUEL

Currently operating reprocessing plants all use variations on the PUREX process. In this process, spent nuclear fuel is chopped and cladding hulls are separated. The chopped fuel assemblies are dissolved in nitric acid, and the solution is prepared with organic flocculating agents and filtration for the extraction process. Extraction of uranium, plutonium, and neptunium is accomplished by tributyl phosphate (TBP) solutions in hydrocarbon dissolvent. Uranium and plutonium products of the process are almost entirely free of fission product. Uranium and plutonium are separated from each other to better than 1 part in 7×10^5, with waste losses of uranium, plutonium, and neptunium less than or equal to 0.01 percent, 0.025 percent, and 0.5 percent, respectively (Myasoedov, 2007). In addition to plants built for separating weapons plutonium, large plants of this kind are separating plutonium from civilian fuel in France, Japan, Russia, and the United Kingdom; two small plants are operating in India; and China has recently built a pilot plant. The Russian plant, RT-1, located at the Mayak Production Association in the town of Ozersk, was launched in 1976, and processes fuel from both propulsion and power reactors.

The United States and Russia have accumulated large stocks of spent nuclear fuel. The United States every year adds 2,000 MTHM of spent nuclear fuel to its stored inventory, which reached 58,000 MTHM in 2007. By 2016, the inventory will be about 77,000 MT, which is over the 63,000 MTHM legal limit for commercial power-reactor waste to be disposed in the first high-level waste repository in the United States.[17] The Russian Federation adds 700 MTHM of spent nuclear fuel each year to its stores, which now are at about 16,000 MT. By 2016, Russia anticipates it will have more than 25,000 MTHM of spent fuel in storage. To develop options for these stocks of spent fuel and for future fuel cycles, several research programs have examined partitioning of key radionuclides to improve the overall performance of the repository. Development of improved processes for extracting key radionuclides from spent fuel and of improved reactor and fuel technologies would be needed to achieve the ambitious goals for reducing the repository burden that GNEP and some other national programs have set.

In both cases the partitioning of radionuclides has the potential to make changes in waste streams that could improve repository performance. Most important among the ones relevant to the fuel cycle options considered here are improved waste forms and reduced total actinide

[17] The technical or geologic limit at the proposed site, Yucca Mountain, is expected to be larger than the legal limit. The Electric Power Research Institute (EPRI) has estimated the technical capacity to be four to nine times greater (EPRI, 2007).

content, which lowers the heat loading and the long-term radiotoxicity in the repository. The mobile radiotoxicity (which is more relevant than the radiotoxicity itself and is repository dependent) could be lowered in the context of Yucca Mountain if actinides are burned.[18] The Yucca Mountain Program has, however, stated that the repository will meet its licensing requirements without reductions of radiotoxicity within the legal capacity of the repository.

If heat load is a limiting factor in repository capacity,[19] then reducing the heat load in the waste streams would enable a country to dispose of the waste from more nuclear electricity generation within a repository of fixed capacity (though most countries are planning on repository sites that would be readily expandable). Given the difficulties already encountered in siting and opening a repository, there may be a significant benefit in extending a repository's capacity. How much of a difference recycling can make depends very much on the details of the burn-up, the waste streams, the waste forms, and the specific repository design and environment, so only a scenario-based approach to analysis works, and right now there is not enough information to know which scenarios are most likely. This approach to increasing repository capacity or reducing repository hazards, however, entails a trade-off with the siting and hazards associated with additional facilities for handling and processing the materials aboveground in the closed fuel cycle.

B6. Are new laws and/or regulations required for either the U.S. or the Russian approach to the internationalization of the fuel cycle? Will either approach require any existing laws or regulations to be repealed or changed?

As noted in Section A8, there are many laws, regulations, and legal instruments that would need to be revised to reduce "road blocks" to proliferation threat reduction. Key among those is bringing into force a civilian nuclear cooperation agreement (known as a 123 agreement for the relevant section of the U.S. Atomic Energy Act (AEA); see Box 3.5) with Russia and any other nation that is critical to the successful implementation of international fuel cycles involving transfer of spent nuclear fuel. Because a substantial fraction of the world's stock is U.S.-obligated fuel, which cannot be transferred to another party without both a 123 agreement and U.S. approval, any international scheme for spent fuel management is necessarily limited by the lack of a civilian nuclear cooperation agreement with the United States. Such an agreement would be necessary for a future international center for spent fuel management to be able to

---

[18] For many years, analyses of the proposed repository at Yucca Mountain have identified neptunium-237 as the dominant contributor to potential dose from groundwater consumption in long time frames (beyond several tens of thousands of years), with technetium-99, carbon-14, and iodine-129 dominating in earlier time frames. However, estimates of actinide contributions to potential dose in the long term have been reduced very recently (DOE, 2008b, pg. 5-6) because the U.S. Department of Energy applied revised International Commission on Radiological Protection (ICRP) weighting factors for calculation of individual doses (ICRP, 2001). Now "[t]he estimated mean annual individual dose [beyond 10,000 years] at the [reasonably maximally exposed individual] location would consist of approximately 30 percent from plutonium-242, about 20 percent from each of iodine-129 and neptunium-237, about 15 percent from radium-226, and about 8 percent from technetium-99." (DOE, 2008, p. 5-30; ICRP, 2001)

[19] Some argue that long-term heat load need not be a limiting factor for repositories, because repositories in the saturated zone (those located below the underground water table) have abundant water to absorb and carry away heat, and repositories in the unsaturated zone (those located above the water table) can be left open with air circulating to remove heat. However, all repository designs have some heat considerations. For example, some repositories in saturated zones use bentonite clay to inhibit water flow past waste packages and retard contaminant transport from the waste; but the clay properties worsen as the clay temperature rises (see, e.g., Neall, 2008).

operate effectively in Russia. Politically the United States is unlikely to be able to take back spent fuel itself for many years to come.[20]

[20] Under U.S. law, such take-backs would require congressional approval, though they are not prohibited in principle; such approval is unlikely to be forthcoming, except in special cases, such as the ongoing return of irradiated research reactor fuel, which is part of a program to reduce proliferation risks by eliminating highly enriched uranium (HEU) from as many research reactors as possible.

BOX 3.5
NUCLEAR COOPERATION WITH THE UNITED STATES: AGREEMENTS ON PEACEFUL USES OF NUCLEAR WEAPONS BETWEEN THE UNITED STATES AND RUSSIA

U.S. Atomic Energy Act of 1954, Section 123

Significant nuclear exports from the United States are only legally permitted under Section 123 of the U.S. Atomic Energy Act (AEA) of 1954 as amended, 42 U.S.C., Section 2153, in accordance with an agreement for peaceful nuclear cooperation with the recipient.* Such agreements are frequently referred to as 123 agreements.† Exports deemed significant include power reactors, research reactors, nuclear source material (including reactor fuel), and four major components of reactors (pressure vessels, fuel charging and discharging machines, complete control rod drive units, and primary coolant pumps). A 123 agreement between the United States and another country establishes a framework for exports and cooperation, but does not obligate the United States to provide nuclear exports to the recipient country, or to engage in specific cooperative activities.

Section 123 of the AEA requires that the following key conditions and requirements be included in a U.S. agreement for peaceful nuclear cooperation:[a]

- a guarantee by the cooperating party that safeguards will be maintained with respect to all nuclear materials and equipment transferred, and with respect to all special nuclear material used in or produced through the use of such nuclear materials and equipment

- a guarantee that no nuclear materials and equipment or sensitive nuclear technology will be used for any nuclear explosive device, or for research on or development of any nuclear explosive device, or for any other military purpose

- except in agreements with nuclear weapon states, a stipulation that the United States shall have the right to require the return of any nuclear materials and equipment transferred to the recipient country and any special nuclear material produced through the use thereof if the cooperating party detonates a nuclear explosive device or terminates or abrogates an agreement providing for International Atomic Energy Agency safeguards

- a guarantee that any material or any restricted data transferred pursuant to the agreement and, except in specific cases, any production or utilization facility transferred pursuant to the agreement or any special nuclear material produced through the use of any such facility or through the use of any material transferred pursuant to the agreement, will not be transferred to unauthorized persons or beyond the jurisdiction or control of the cooperating party without the consent of the United States

- a guarantee that adequate physical security will be maintained with respect to any nuclear material transferred and with respect to any special nuclear material used in or produced through the use of any material, production facility, or utilization facility transferred

- a guarantee that no material transferred and no material used in or produced through the use of any material, production facility, or utilization facility transferred will be reprocessed, enriched, or otherwise altered in form or content without the prior approval of the United States

- a guarantee that no plutonium, no uranium-233, and no uranium enriched to greater than 20 percent in the isotope 235, transferred pursuant to the agreement or recovered from any source or special nuclear material so transferred or from any source or special nuclear material used in any production facility or utilization facility transferred pursuant to the agreement, will be stored in any facility that has not been approved in advance by the United States

- a guarantee that any special nuclear material, production facility, or utilization facility produced or constructed under the jurisdiction of the cooperating party by or through the use of any sensitive nuclear technology transferred will be subject to all the requirements specified above

In addition to the full list of specified requirements, it is not uncommon for 123 agreements to also apply reciprocal nonproliferation conditions, assurances, and controls. Although not required by U.S. law, the United States may accept the obligations contained in the agreement on a reciprocal basis should it import materials or equipment from the cooperating party.

Proposed 123 agreements are to be negotiated by the secretary of state, "with the technical assistance and concurrence of the secretary of energy and after consultation with the (Nuclear Regulatory) Commission." Following negotiations, the proposed agreement is to be submitted to the President for review. The President must submit an agreement for cooperation to Congress for a statutory review period of 90 days continuous session; however, the actual review period may extend over several more months, depending on the congressional schedule. The Russian Federation and the United States signed an agreement on nuclear energy cooperation, which the United States considers a 123 agreement, on May 6, 2008.

Approval and enactment of a 123 agreement does not require the approval of Congress, but Congress may enact legislation to disapprove the agreement. If there is no prohibitory legislation, an agreement may be brought into force following the close of the congressional review period. Once an agreement for cooperation has been brought into force, exports made under the agreement require a license from the U.S. Nuclear Regulatory Commission and must be consistent with other sections of the AEA (Sections 127 and 128) pertaining to the U.S. nuclear export criteria.

* Atomic Energy Act, 1954.
† Currently the United States has 123 agreements with 19 individual countries plus Taiwan and 2 international organizations, the International Atomic Energy Agency and Euratom (which includes 27 individual countries).
[a] For a comprehensive list of requirements, see Atomic Energy Act, 1954.

The United States already has such agreements with[21] Argentina, Australia, Bangladesh, Brazil, Canada, China, Colombia, Egypt, the European Atomic Energy Community (Euratom),[22] Indonesia, the International Atomic Energy Agency (IAEA), Japan, Kazakhstan, the Republic of Korea, Morocco, Norway, South Africa, Switzerland, Taiwan,[23] and Thailand. As noted above,

[21] Information about current agreements is taken from "123 Agreements for Peaceful Cooperation" an information sheet available (as of August 31, 2008) at http://nnsa.energy.gov/nuclear_nonproliferation /123_agreements_peaceful_cooperation.htm.

[22] Euratom comprises the following member states: Austria, Belgium, Bulgaria, Cyprus, the Czech Republic, Denmark, Estonia, Finland, France, Germany, Greece, Hungary, Ireland, Italy, Latvia, Lithuania, Luxembourg, Malta, the Netherlands, Poland, Portugal, Romania, Slovakia, Slovenia, Spain, Sweden, and the United Kingdom.

[23] Pursuant to Section 6 of the Taiwan Relations Act, P.L. 96-8, 93 Stat. 14, and Executive Order 12143, 44 F.R. 37191, all agreements concluded with the Taiwan authorities prior to January 1, 1979, are administered on a

the United States and Russia negotiated such an agreement, but the U.S. Congress did not vote on it and on September 8, 2008, President George W. Bush withdrew it from consideration (Rice, 2008).

The United States and Russia are leaders in nuclear technology. The vast majority of nuclear energy technology currently developed worldwide was developed in Russia and the United States. These two nations also have the most developed technologies and technical capabilities to support nuclear nonproliferation. Both have invested a great deal of time and energy in developing concepts to advance the concept of a safer, more secure international nuclear fuel cycle program. Russia and the United States are able to conduct civilian nuclear energy cooperation with the other leaders in nuclear energy, but not with each other, and the lack of a U.S.-Russian agreement restricts those partners' cooperation on nuclear energy with Russia and the United States. It is difficult to see how such an international program could move forward without the active participation and (probably) cooperation of these two countries. But the appropriate mechanism must be in place to allow this kind of cooperation.

Other considerations beyond the scope of this study will factor in to the decisions by the U.S. President and Congress whether or not to bring the signed agreement into force (Einhorn et al., 2008; Alvarez, 2008). The joint committees recognize that it is unlikely that the U.S. government will bring the agreement into force in an environment of worsening relations between the United States and Russia. It is the joint committees' hope that current disagreements that have recently emerged will not interfere with the United States and Russia working together toward their common goal of inhibiting nuclear weapons proliferation as nuclear energy use grows across the world.

Finding 12

The United States and the Russian Federation have signed an agreement on peaceful nuclear cooperation, but it must still be allowed to come into force. The lack of a U.S.-Russian agreement in force is interfering with joint efforts to reduce proliferation. U.S.-Russian cooperation on nuclear energy technology that involved the transfer of nuclear materials, major elements of reactor designs and technology, or major elements of fuel cycle designs and technology from the United States to Russia is only possible under a bilateral agreement on nuclear cooperation (called a 123 agreement in the United States). The expanded cooperation in nuclear energy research and development and commercial implementation that such a bilateral cooperation could make possible could serve both countries' interests in expanding the use of nuclear energy while meeting safety, security, and nonproliferation objectives. Approval of such an agreement could help establish an atmosphere of cooperation that will strengthen prospects from cooperative international approaches to the fuel cycle and other nonproliferation problems. In particular, under U.S. law, international fuel cycle approaches that involved take-back of fuel to Russia (the only country that yet has a legal structure in place for such take-back) would have to exclude all U.S.-obligated material until a civil cooperation agreement had been put in place.

---

nongovernmental basis by the American Institute in Taiwan, a nonprofit District of Columbia corporation, and continuation of any official relationship with Taiwan.

LIST OF ACRONYMNS

AEA	U.S. Atomic Energy Act
AEC	U.S. Atomic Energy Commission
AECC	Angarsk Electrolysis Chemical Complex (Russia)
ANPP	Armenian Nuclear Power Plant
CANDU	Canadian deuterium-uranium reactors
CAS	Committee on Assurances of Supply (IAEA)
EBR-II	Experimental Breeder Reactor II Program (U.S.)
EFEI	Experimental Fuel Element Installation (Indonesia)
ETC	Enrichment Technology Company (Urenco)
FEPI	Fuel Element Production Installation (Indonesia)
GIF	Generation-IV International Forum
GNEP	Global Nuclear Energy Partnership (U.S.)
HEU	highly enriched uranium
HLW	high level waste
HTGR	high-temperature gas-cooled reactor
IAEA	International Atomic Energy Agency
INFA	International Nuclear Fuel Authority (U.S.)
INFCE	International Nuclear Fuel Cycle Evaluation
INRNE BAS	Institute for Nuclear Research and Nuclear Energy of the Bulgarian Academy of Sciences
ITER	International Thermonuclear Experimental Reactor (U.S.)
IUEC	International Uranium Enrichment Center (Russia)
LEU	low-enriched uranium
LCGP	Least Cost Generation Plan (Armenia)
LWR	light-water reactor
MOX	mixed-oxide fuel
NEF	National Enrichment Facility (U.S.)
NNPA	U.S. Nuclear Nonproliferation Act of 1978
NNWS	nonnuclear-weapons states
NPP	nuclear power plant
NPT	Treaty on the Non-Proliferation of Nuclear Weapons
NSG	Nuclear Suppliers Group
OECD	Organisation for Economic Co-operation and Development
PWR	pressurized water reactor
RANF	Reliable Access to Nuclear Fuel
RIAR	Research Institute of Atomic Reactors (Russia)
SWU	separative work units
TBP	tributyl phosphate solutions
THTR	thorium high-temperature reactor
TRU	transuranics
USAID	U.S. Agency for International Development

REFERENCES

Alvarez, Robert. 2008. U.S.-Russian Nuclear Agreement Raises Serious Concerns. *Bulletin of the Atomic Scientists*. 16 June.

Ansolabehere, Stephen, John Deutch, Michael Driscoll, Paul E. Gray, John P. Holdren, Paul L. Joskow, Richard K. Lester, Ernest J. Moniz, Neil E. Todreas, Eric S. Beckjord, Nathan Hottle, Christopher Jones, and Etienne Parent. 2003. The Future of Nuclear Power: An Interdisciplinary MIT Study. Massachusetts Institute of Technology.

APS (American Physical Society). 2008. Readiness of the U.S. Nuclear Workforce for 21st Century Challenges. College Park, MD: APS. Accessed at www.aps.org/policy/reports/popa-reports/upload/Nuclear-Readiness-Report-FINAL-2.pdf on December 13, 2008.

APS. 2005. Nuclear Power and Proliferation Resistance: Securing Benefits, Limiting Risk. College Park, MD: APS. Accessed at www.aps.org/policy/reports/popa-reports/proliferation-resistance/upload/proliferation.pdf on December 13, 2008.

Baruch, Bernard. 1946. The Baruch Plan. Presentation to the United Nations Atomic Energy Commission, New York, June 14.

Boston Consulting Group. 2006. Economic Assessment of Used Nuclear Fuel Management in the United States. Accessed at www.bcg.com/impact_expertise/publications/files/Economic_Assessment_Used_Nuclear_Fuel_Mgmt_US_Jul2006.pdf on November 29, 2008.

Bunn, Matthew. 2007. Proliferation-Resistance (and Terror-Resistance) of Nuclear Energy Systems. Unpublished lecture, November 20. Accessed at http://belfercenter.ksg.harvard.edu/files/uploads/bunn_proliferation_resistance_lecture.pdf on March 22, 2008.

Bunn, Matthew, Steve Fetter, John Holdren, and Bob van der Zwaan. 2003. *The Economics of Reprocessing vs. Direct Disposal of Spent Nuclear Fuel.* Cambridge, Mass.: Managing the Atom Project, Harvard University.

Bunn, Matthew, John P. Holdren, Allison Macfarlane, Susan E. Pickett, Atsuyuki Suzuki, Tatsujiro Suzuki and Jennifer Weeks. 2001. *Interim Storage of Spent Nuclear Fuel. A Safe, Flexible, and Cost-Effective Near-Term Approach to Spent Fuel Management.* Cambridge, Mass.: Managing the Atom Project, Harvard University and Project on Sociotechnics of Nuclear Energy, University of Tokyo.

Bush, George W. 2004. Remarks by the President on Weapons of Mass Destruction Proliferation at National Defense University. February 11. Accessed at www.whitehouse.gov/news/releases/2004/02/20040211-4.html on November 27, 2008.

Cabrera-Palmer, B. and G. Rothwell. 2008. Why Brazil is Enriching Uranium. *Elsevier's Energy Policy* 36: 2570-2577.

CBO (Congressional Budget Office). 2008. Nuclear Power's Role in Generating Electricity. Washington, DC: CBO. Accessed at www.cbo.gov/ftpdocs/91xx/doc9133/toc.htm on December 13, 2008.

Collins, Emory D., Guillermo D. DelCul, John P. Renier, and Barry B. Spencer. 2007. Preliminary Multicycle Transuranic Actinide Partitioning-Transmutation Studies. ORNL/TM-2007/24. Oak Ridge, TN: U.S. DOE. Accessed at www.gnep.energy.gov/peis/references/RM163_Collinsetal_2007.pdf on November 29, 2008.

DOE (Department of Energy). 2008a. Evaluation of Existing Department of Energy (DOE) Facilities to Support the Advanced Fuel Cycle (AFCF) Mission, GNEP-AFCF-PMO-AI-EA-2008-000271. Washington, DC: U.S. DOE.

DOE. 2008b. Final Supplemental Environmental Impact Statement for a Geologic Repository for the Disposal of Spent Nuclear Fuel and High-Level Radioactive Waste at Yucca Mountain, Nye County, Nevada. DOE/EIS-0250F-S1.

DOE. 2002. A Technology Roadmap for Generation IV Nuclear Energy Systems, GIF-002-00. Washington, D.C.: U.S. DOE. Accessed at http://gif.inel.gov/roadmap/pdfs/gen_iv_roadmap.pdf on December 13, 2008.

DOE. 1997. Nonproliferation and Arms Control Assessment of Weapons-Usable Fissile Material Storage and Excess Plutonium Disposition Alternatives. DOE/NN-0007. Washington, DC: U.S. DOE.

DOE. Undated. Spent Nuclear Fuel Transportation. Washington, DC: DOE. Accessed at www.ocrwm.doe.gov/transport/pdf/snf_trans.pdf on November 26, 2008.

Einhorn, R., R. Gottemoeller, F. McGoldrick, D. Poneman, and J. Wolfsthal. 2008. The U.S.-Russia Civil Nuclear Agreement: A Framework for Cooperation. Washington, DC: Center for Strategic and International Studies. Accessed at www.carnegieendowment.org/static/npp/reports/csis_us-russia_08.pdf on December 13, 2008.

ElBaradei, Mohamed. 2008. Reviving Nuclear Disarmament. Speech at Conference on Achieving the Vision of a World Free of Nuclear Weapons, Oslo, February 26.

ElBaradei, Mohamed. 2007. Statement to the Sixty-Second Regular Session of the United Nations General Assembly, Vienna, Austria, October 29, 2007. Accessed at www.iaea.org/NewsCenter/Statements/2007/ebsp2007n018.html on November 26, 2008.

ElBaradei, Mohamed. 2003. Towards a Safer World. *The Economist* October 18.

Electric Power Research Institute. 2007. Program on Technology Innovation: Room at the Mountain. Analysis of the Maximum Disposal Capacity for Commercial Spent Nuclear Fuel in a Yucca Mountain Repository. Technical Report 1015046.

Finck, P. J. 2006. Technologies for Advanced Fuel Cycles. Presentation to the Joint National Academies'-Russian Academy of Sciences' Committees on Internationalization of the Civilian Nuclear Fuel Cycle, Washington, DC, October 17.

Forden, Geoffrey and John Thompson. 2006. (Revised 2007). Iran as a Pioneer Case for Multilateral Nuclear Arrangements. Cambridge, Mass.: Massachusetts Institute of Technology. Accessed at mit.edu/stgs/irancrisis.html on April 28, 2008.

Forwood, Martin. 2008. The Legacy of Reprocessing in the United Kingdom, Research Report No. 5. Princeton: International Panel on Fissile Materials. Accessed at www.nuwinfo.se/files/forwood200807reprocessing-uk-rr05.pdf on December 13, 2008.

Fthenakis, V. M. and H. C. Kim. 2007. Greenhouse-gas Emissions from Solar Electric- and Nuclear Power: A Life-cycle Study. *Energy Policy* 35 (4): 2549-2557.

Galperin, A. and M. Todosow. 2001. Thorium Based Fuel Designed to Reduce the Proliferation Potential and Waste Disposal Requirements of Light Water Reactors. Paper Presented at the Workshop on Advanced Reactors With Innovative Fuels. Chester, UK, 22-24 October. Accessed at www.nea.fr/html/science/meetings/arwif2001/57.pdf on December 13, 2008.

Gat, U., S. M. Crosley, and R. L. Gay. 1993. Molten Salt Treatment to Minimize and Optimize Waste. Washington, DC: DOE. Accessed at www.osti.gov/energycitations/servlets/purl/10173195-gPLIGd/10173195.pdf on December 19, 2008.

Glaser, A. 2006. On the Proliferation Potential of Uranium Fuel for Research Reactors at Various Enrichment Levels. *Science and Global Security* 14:1–24.

Harrison Selig S. 1996. Japan's Nuclear Future: The Plutonium Debate and East Asian Security. Washington, DC: Carnegie Endowment for International Peace.

IAEA (International Atomic Energy Agency). 2008. Commission of Eminent Persons, Reinforcing the Global Nuclear Order for Peace and Prosperity: The Role of the IAEA to 2020 and Beyond.

IAEA. 2007a. Communication received from the Resident Representative of the Russian Federation to the IAEA on the Establishment, Structure and Operation of the International Uranium Enrichment Centre. INFCIRC/708. Accessed at www.iaea.org/Publications/Documents/Infcircs/2007/infcirc708.pdf on November 29, 2008.

IAEA. 2007b. Energy, Electricity and Nuclear Power Estimates for the Period up to 2030. Reference Data Series No. 1, 2007 Edition.

IAEA. 2007c. Multilateralizing the Nuclear Fuel Cycle. INFCIRC/704.

IAEA. 2007d. Possible New Framework for the Utilization of Nuclear Energy: Options for Assurance of Supply of Nuclear Fuel. GOV/INF/2007/11.

IAEA. 2006a. Communication received from the Resident Representative of the Russian Federation to the Agency transmitting the text of the Statement of the President of the Russian Federation on the Peaceful Use of Nuclear Energy. INFCIRC/667. Accessed at www.iaea.org/Publications/Documents/Infcircs/2006/infcirc667.pdf on November 29, 2008.

IAEA. 2006b. Global Uranium Resources to Meet Projected Demand. Accessed at www.iaea.org/NewsCenter/News/2006/uranium_resources.html on December 19, 2008.

IAEA. 2005a. Innovative Small and Medium Sized Reactors: Design Features, Safety Approaches and Research and Development Trends. IAEA-TECDOC-1451. Accessed at www-pub.iaea.org/MTCD/publications/PDF/TE_1451_web.pdf on December 13, 2008.

IAEA. 2005b. Multilateral Nuclear Approaches to the Nuclear Fuel Cycle. INFCIRC/640. Accessed at www.iaea.org/Publications/Documents/Infcircs/2005/infcirc640.pdf on December 13, 2008.

IAEA. 1980. Assurances of Long-Term Supply of Technology, Fuel and Heavy Water and Services in the Interest of National Needs Consistent with Non-Proliferation.

ICRP (International Commission on Radiological Protection). 2001. The ICRP Database of Dose Coefficients: Workers and Members of the Public. ICRP Publication 72. New York: Elsevier.

IEA (International Energy Agency). 2007. *IEA Energy Technology Essentials: Nuclear Power.* Accessed at www.iea.org/textbase/techno/essentials4.pdf on November 29, 2008.

(ISAB) International Security Advisory Board. 2008. Report on Proliferation Implications of the Global Expansion of Civil Nuclear Power. Washington, DC: U.S. Department of State. Accessed at www.ne.doe.gov/pdfFiles/rpt_ISAB_GlobalExpansionofCivilNuclearPower_Apr2008.pdf on December 13, 2008.

Kidd, Steve. 2007. Russia: Where's It Heading in Nuclear? *Nuclear Engineering International Magazine* 14 May. Accessed at www.neimagazine.com/story.asp?storyCode=2044327 on December 13, 2008.

Kirienko, S. V. 2007. Report to the Plenary Session of the Fifty-first IAEA General Conference, Vienna, Austria, September 17-21.

Kuliasha, Michael A. and Alexander Zucker. 1992. Technologies for a Greenhouse-Constrained Society. Boca Raton, FL: CRC Press.

Laidler, J. J. 2007. GNEP Spent Fuel Processing; Waste Streams and Disposition Options. Presentation to the Nuclear Waste Technical Review Board, Washington, DC, May 15.

Lasky-Fink, Jessica. 2008. Brazil, Argentina Pursue Nuclear Cooperation. *Arms Control Today* May.

Meckoni, V., R J. Catlin, and L. L. Bennett. 1977. Regional Nuclear Fuel Cycle Centres. IAEA Study Project. International Conference on Nuclear Power and its Fuel Cycle, Salzburg, Austria.

Meier, Oliver. 2006. Iran and Foreign Enrichment: A Troubled Model. *Arms Control Today* January/February.

Milhollin, Gary. 1987. Stopping the Indian Bomb. *American Journal of International Law* 81: 593.

Myasoedov, Boris. 2007. Wet Methods of Spent Nuclear Fuel Reprocessing: Current Status and Future Prospects. Presentation to the Joint National Academies'-Russian Academy of Sciences' Committees on Internationalization of the Civilian Nuclear Fuel Cycle, Vienna, Austria, April 22.

National Academy of Sciences. 1994. *Management and Disposition of Excess Weapons Plutonium.* Washington, DC: National Academy Press.

Neall, Fiona. 2008. Thermal Drivers of Spent Fuel and High Level Waste Repository Design in Hard Rock. Unpublished presentation.

New York Times. April 16, 1981. Westinghouse Ends Last Uranium Suit.

Norman, Colin. 1986. Uranium Enrichment's $7 Billion Uncertainty. *Science* 232: 4748: 312-313.
NRC (National Research Council). 2007. *Review of DOE's Nuclear Energy Research and Development Program.* Washington, DC: The National Academies Press.
NRC. 2006. *Going the Distance? The Safe Transport of Spent Nuclear Fuel and High-Level Radioactive Waste in the United States.* Washington, DC: The National Academies Press.
NRC. 2005. *An International Spent Nuclear Fuel Storage Facility.* Washington, DC: The National Academies Press.
NRC. 2003. *End Points for Spent Nuclear Fuel and High-Level Radioactive Waste in the United States and Russia.* Washington, DC: The National Academies Press.
NRC. 2002. *Making the Nation Safer: The Role of Science and Technology in Countering Terrorism.* Washington, DC: The National Academies Press.
NRC. 2000. *The Spent-Fuel Standard for Disposition of Excess Weapon Plutonium: Application to Current DOE Options.* Washington, DC: National Academies Press.
Nuclear Energy Agency. 2006. Evaluation Methodology for Proliferation Resistance and Physical Protection of Generation IV Nuclear Energy Systems, Revision 5. GIF/PRPPWG/2006/005. Paris: NEA for the Generation IV International Forum. Accessed at www.gen-4.org/Technology/horizontal/PRPPEM.pdf on December 13, 2008.
Nuclear Power Today. 2005. September 30, 2005. Poll Reveals Russian Support for New Nuclear Projects.
OECD (Organisation for Economic Co-operation and Development)/IAEA. 2008. Uranium 2007: Resources, Production and Demand. *Red Book.*
OECD/IAEA. 2006. Uranium 2005—Resources, Production and Demand.
Overview of Galena's Proposed Approach to Licensing a 4S Nuclear Reactor Based Power Generation Facility. Rev 02, 3/12/2007. Galena, Alaska. City Government. Accessed at www.roe.com/pdfs/technical/Galena/Overview%20Whitepaper%20Rev02.pdf on March 29, 2008.
Putin, Vladimir V. 2000. Address to the United Nations Millennium Summit, September 6, 2000. Accessed at www.un.org/millennium/webcast/statements/russia.htm on November 27, 2008.
Rice, Condoleezza. 2008. Statement by Secretary Condoleezza Rice on U.S.-Russia 123 Agreement, Washington, DC, September 8. Accessed at www.state.gov/secretary/rm/2008/09/109256.htm on December 13, 2008.
Rickover, Hyman. 1970. Paper Reactors, Real Reactors. Atomic Energy Committee Authorizing Legislation: Hearings Before the Joint Committee on Atomic Energy. June 5, 1953
Rockwell, Theodore. 2002. *Rickover Effect: How One Man Made a Difference.* Bloomington, IN: iUniverse, Incorporated
Ruchkin, S. V. and V. Y. Loginov. 2006. Securing the Nuclear Fuel Cycle: What Next? *IAEA Bulletin* 48/1.
Schneider, Erich A. and William C. Sailor. 2008. Long-Term Uranium Supply Estimates. *Nuclear Technology* 162 (June): 379-387.

Schneider, Mycle and Yves Marignac. 2008. Spent Nuclear Fuel Reprocessing in France. Research Report No. 4. Princeton: International Panel on Fissile materials. Accessed at www.fissilematerials.org/ipfm/site_down/rr04.pdf on December 13, 2008.

Selig S. Harrison, ed. 1996. Japan's Nuclear Future: The Plutonium Debate and East Asian Security. Washington, DC: Carnegie Endowment for International Peace.

Shaker, Mohamed I. 2007. The Internationalization of the Nuclear Fuel Cycle: An Arab Perspective. Paper presented at the Workshop on the Internationalization of the Nuclear Fuel Cycle, Vienna, Austria. April 23.

Shropshire, D. E., K. A. Williams, W. B. Boore, J. D. Smith, B. W. Dixon, M. Dunzik-Gougar, R. D. Adams, D. Gombert, and E. Schneider. 2008. Advanced Fuel Cycle Cost Basis. INL/EXT-07-12107. Idaho Falls: Idaho National Laboratory.

U.S. Congress. 1954. Atomic Energy Act. 42 U.S.C. § 2011 et seq. Accessed at www.nrc.gov/reading-rm/doc-collections/nuregs/staff/sr0980/ml022200075-vol1.pdf#pagemode=bookmarks&page=14 on April 2, 2008.

U.S. GAO (Government Accounting Office). 1978. Quick and Secret Construction of Plutonium Reprocessing Plants: A Way to Nuclear Weapons Proliferation? GAO EMD-78-104. Washington, DC: U.S. GAO. Accessed at http://archive.gao.gov/f0902c/107377.pdf on December 13, 2008.

U.S. Mission to International Organizations in Vienna. Undated. Fact Sheet on U.S. HEU for a Nuclear Fuel Reserve. Accessed at http://vienna.usmission.gov/np_nuclear.html on December 13, 2008.

U.S. Nuclear Regulatory Commission. 2005. Transportation of Spent Fuel and Radioactive Materials. Washington, DC: Office of Public Affairs. Accessed at www.nrc.gov/reading-rm/doc-collections/fact-sheets/transport-spenfuel-radiomats-bg.html on December 13, 2008.

van Doren, Charles N. 1983. Nuclear Supply and Non-Proliferation: The IAEA Committee on Assurances of Supply. Washington, DC: Congressional Research Service.

WNA (World Nuclear Association). 2008. Thorium Fact Sheet. July. Accessed at http:www.world-nuclear.org/info on December 13, 2008.

WNA. 2006. The Global Nuclear Fuel Market: Supply and Demand 2007-2030. London.

APPENDIX A

STATEMENT OF TASK

This joint study by the U.S. National Academies and the Russian Academy of Sciences (NAS and RAS) will provide an assessment of the technical, economic, legal/regulatory, and nonproliferation criteria necessary for the implementation of an international civilian nuclear fuel cycle. The study is not intended to be a comprehensive treatment of the topics listed, but rather a high-level, first cut at these complex issues. Specifically, the proposed NAS-RAS joint study will address the primary issues and questions listed below under headings A and B. The secondary issues and questions will be addressed to the extent that budget and time permit.

A. Providing fuel services to countries that already have Light Water Reactors (LWRs) or would be interested in constructing LWRs if they did not have to develop the entire fuel cycle.

Primary Issues:

1. Is it feasible and effective to establish international fuel supply centers as an incentive for countries not to develop indigenous enrichment facilities?
2. What are the advantages and disadvantages (if any) of establishing international centers for: sending and receiving back fuel? Training personnel? Manufacturing fuel?
3. Who should own the nuclear material and the fuel in such arrangements?
4. Should the international facilities be owned by governments or could private companies own some or all of the facilities?

Secondary Issues:

5. What regulatory requirements should be in place in the receiving country to provide assurance of safety and safeguards?
6. What level of technical personnel are needed, in terms of training and in terms of numbers, to provide adequate confidence that the countries receiving fuel can safely and securely operate their reactor(s)?
7. What should be the role of the International Atomic Energy Agency in overseeing the transfer, use, and/or return of fuel?
8. What changes in laws and regulations in the countries sending, consuming, and receiving spent fuel would be required to implement this concept?

B. Fuel Regeneration Options to Support an International Nuclear Fuel Cycle.

Primary Issues:

1. Compare the uranium recovery by extraction plus (UREX+), the plutonium and uranium recovery by extraction (PUREX) process, and other processes being considered by the Russian Federal Agency for Atomic Energy for separation of fissile and other materials from spent or irradiated nuclear fuel. Consider the resulting waste streams and what can and should be done with these waste streams.
2. Compare the burn up and the number of cycles needed to reach an acceptable level of destruction of actinides in the conceptual advanced burner reactor proposed in the U.S. Global Nuclear Energy Partnership (GNEP) and in the Russian BN-600 and BN-800 reactors.
3. What impact could new technologies have on these proposals?

Secondary Issues:

4. Compare the fuel to be produced from the processes examined in (1) for use in appropriate reactors (LWRs, High Temperature Gas Cooled Reactors, and fast reactors). What are the advantages and disadvantages of each type of fuel?
5. Compare the repository requirements for the waste produced by the processes proposed in the GNEP concept with that from a system based on PUREX and one based on Russian plans.
6. Are new laws and/or regulations required for either the U.S. or the Russian approach to the internationalization of the fuel cycle? Will either approach require any existing laws or regulations to be repealed or changed?

Because the scale of the full study task is large and the details of proposed fuel cycle strategies are in flux, the study will be carried out in two phases. In Phase I, the joint committees will identify distinct strategies that represent the range of fuel cycle options and gather the key technical and legal/regulatory and other information needed to analyze those options. This information-gathering stage will culminate with an international workshop. In Phase II, the joint committees will carry out the analysis and offer consensus findings and recommendations in a final report on the criteria necessary to achieve an international fuel cycle beneficial for suppliers and consumers alike and supportive of international nonproliferation efforts. The final report will be subject to the NAS/National Research Council report review process, and will be sent to Russian reviewers as well as to U.S. reviewers.

APPENDIX B

WORKSHOP ON INTERNATIONALIZATION OF THE NUCLEAR FUEL CYCLE

Convened by
the U.S. National Academies (NAS) and the
Russian Academy of Sciences (RAS)
With the support of the International Atomic Energy Agency
Vienna, Austria

Summary by Rita Guenther, Marc Humphrey, and Micah Lowenthal

WORKSHOP – DAY 1[1]

Monday, April 23, 2007

Welcome from Tariq Rauf and Alan McDonald, IAEA, and Co-chairs Dr. John Ahearne and Academician Boris Myasoedov

Alan McDonald of the International Atomic Energy Agency (IAEA) welcomed participants and noted that workshop deliberations would have some influence on the June 2007 IAEA Board of Governor's meeting. John Ahearne, chair of the National Academies' committee, observed that the increased interest in nuclear energy across the world was one motivation for this workshop because the increased use of nuclear energy may lead to the spread of enrichment and reprocessing technology, and therefore increased risk of nuclear proliferation.

Ahearne went on to note that there have been several conferences on possible options for guaranteed nuclear fuel supply, and the joint NAS/RAS committees have been examining these various options. The objective of the workshop is to hear from voices outside of the group that has dominated discussions, in particular from voices of experts from key countries, although each participant is acting in a personal capacity and not as a representative of his country. The key questions of the workshop were: How can we increase access to nuclear power? How can we do so while reducing the proliferation risk?

Boris Myasoedov, acting chair of the Russian Academy of Sciences' committee, noted that energy is essential to human development, but fossil resources are limited, even in Russia. As a result, interest in nuclear power is growing quickly. Of course, alternatives such as hydrogen or renewable resources are under investigation. Since the Obninsk reactor was first

[1] For a list of participants, please see the end of this Appendix.

connected to an electrical grid in 1954, nuclear energy has spread to many countries. Facing this increased use, several problems remain including nonproliferation of fissile materials and new, risky technologies. Amidst this fast development, then Russian President Vladmir V. Putin launched an initiative to provide assured access to nuclear services to countries that voluntarily reject the development of some technologies. It offers nuclear resources based on countries having met this requirement, and regardless of political circumstances. Angarsk has been designated as the first international center and in March 2007, the first seminar with IAEA representatives took place in Angarsk at which participants discussed legal aspects of such a center. Recently, Kazakhstan (in a joint presidential meeting) agreed to join this international enrichment center, which will accumulate enriched uranium in gas or solid form and will be the property of the international center. As a commercial enterprise, the center will be open to all countries through intergovernmental agreements. The management and legal aspects of the center are still being discussed. The second stage of the creation of the center will include not only enrichment but the organization of spent fuel return for reprocessing and reuse of the fissile material in nuclear power plants. The U.S. Global Nuclear Energy Partnership (GNEP) proposal also includes reprocessing. Objectives of the workshop are to discuss these various proposals for creating centers, and to hear from those who may wish to use these services.

Tariq Rauf, director of the Office of External Affairs at the IAEA, expressed Director General Mohamed ElBaradei's support for this activity. He noted that the Academies' fuel cycle study is on a longer time frame than the IAEA study of these questions, and that the Director General will provide the IAEA Board of Governors with a new paper on new approaches in June 2007. Rauf noted that to be credible, any plan for assurance of supply must be perceived to be fair and impartial. It was clear at the September 2006 Special Event[2] that no state was ready to give up any rights under the Treaty on the Non-Proliferation of Nuclear Weapons (NPT). In its June paper, the IAEA group needs to reflect this viewpoint, while finding a fair, impartial solution. There is progress, with the visit to Angarsk and the U.S. pledge to downblend over 17 metric tons of highly enriched uranium (HEU) for use in an assured fuel supply arrangement as well as with former Russian President Vladimir V. Putin's initiative, the six-party proposal, and those from the World Nuclear Association,[3] the U.K., and Germany. Rauf expressed his hope that IAEA may be ready for serious movement by the end of 2007.

"Bulgaria and the Internationalization of the Nuclear Fuel Cycle," Jordan Stamenov (Bulgaria)

Bulgaria is a small country with a population of 7.5 million and very limited natural resource reserves. Bulgaria's experience with nuclear technology began in 1955. In 1956, the Joint Institute for Nuclear Research in Dubna, an institution for education and training of nuclear scientists, was created.

The first Bulgarian research reactor began operation in 1961. The IRT-2000, a heterogeneous water-water pool-type reactor (thermal capacity 2 MW), is housed at the Nuclear Scientific and Experimental Center of the Institute for Nuclear Research and Nuclear Energy of the Bulgarian Academy of Sciences (INRNE BAS). From 1990-2002, Bulgaria's first nuclear

[2] For more information on the IAEA Special Event, see http://www.iaea.org/NewsCenter/News/2006/assurancesofsupply.html, accessed on December 13, 2008.

[3] The 2006 World Nuclear Association (WNA) Market Report shows the same growth projection constituted by one BN-1800 coming on line in 2023 and light water reactors for the rest of the growth through 2030 (WNA, 2006).

power plant, the Kozloduy nuclear power plant (NPP), was constructed and commissioned. It comprised 6 units for a total capacity of 3,760 MW, all of first and second generation Soviet design. Fifteen years ago the Kozloduy units 1 to 4 were considered to be "not economically upgradeable" and were closed to meet European Union ascension demands. The latest IAEA inspection, however, confirmed that this "non-upgradeable" definition is no longer true and perhaps it never was. The radiation dose at this plant lies within international standards. They have worked safely for more than 100 reactor-years. Spent fuel from Kozloduy is stored on site then sent to Russia for reprocessing. Bulgaria has no final spent nuclear fuel storage facilities, and uranium mining was ceased 15 years ago.

The main principles of Bulgaria's energy policy include a transparent and unbiased economic environment and investments in energy efficiency. Regionally, Bulgaria is considered to have a well established energy infrastructure, and very good transmission capacities. It has an opportunity to become a leader in the regional energy market. In 2001, it exported 7 GWh of electricity and 14.5 billion cubic meters of natural gas. By 2010, electricity consumption will meet electricity production, minimizing exports. Bulgaria is one of the most energy intensive countries and one of the most energy import-dependent countries in Europe.

Bulgaria has a goal-oriented energy policy, calling for a reliable, efficient, strategic, and socially accepted energy supply. In 2001, nuclear power accounted for 41% of net electricity generation in Bulgaria. The Council of Ministers made a resolution in July 2001 for the reconstruction of the research reactor IRT-2000 into a low-power (200 kW) reactor using low-enriched uranium 235 (LEU) fuel.

On June 24, 2004, the government made the decision to begin construction of an NPP in Belene. Numerous reactor types have been considered and one of the VVER1000 designs was selected. Areva is responsible for construction. Education is one objective for human resource development for the Belene project. Russia used to provide much of the education for nuclear experts in Bulgaria, but now Bulgaria must do much more itself. Another objective is the development of new applications, such as boron neutron capture therapy.

The U.S. and Russian governments signed a contract in May 2004, for cooperation in transferring Russian-produced research reactor nuclear fuel to the Russian Federation, in the frame of two programs: RERTR – Reduced Enrichment of Research and Test Reactors, and RRRFR – Russian Research Reactor Fuel Return.

The IAEA – INRNE BAS initiative has held many meetings with a goal of east-west (VVER-Light Water Reactor) cooperation. In 2007, Bulgaria hosted the 7th International Conference on VVER fuel performance, modeling, and experimental support in cooperation with the IAEA.

Discussion

Gottemoeller: Based on past collaboration between the BAS and Argonne National Lab, what is the view of multilateral cooperation?

Stamenov: We have had scientific connections for more than 35 years. We have also had good cooperation on spent fuel transport.

Gottemoeller: What is your view on the potential feasibility of international arrangements in managing international fuel centers, based on your experience with trilateral agreements (U.S., Russia, Bulgaria)?

Stamenov: Cooperation with Argonne National Laboratory was a natural extension of existing scientific collaborations.

Lowenthal: Were there any important factors leading to the Bulgaria-Russia agreement on a new reactor and fuel?

Stamenov: We have had no problems so far. If there are problems with transport, they are due to transportation schemes. There may be other possibilities, perhaps using ships on the Danube to the Sea. Angarsk has not yet been discussed.

"Perspectives and Policy Options of Nuclear Fuel Cycle Services," Karyono (Indonesia)

Indonesia is an archipelago nation with more than 17,000 islands near the equator, with over 120 million people, and with small uranium reserves. With its high population density and rising electricity consumption, there is now a real need for nuclear power. Indonesia's nuclear program began in the 1970's.

According to "Act Number 10, Year 1997 on Nuclear Energy," the executing body (BATAN) has the right to "undertake the nuclear fuel cycle services that could generate the nuclear and common industries." The main objectives of the nuclear energy program are as follows:

1. short term – statement of nuclear options for long-term planning
2. medium and long term – science and technology foundation, operation of first NPP on the Java-Bali grid, and data collection on uranium reserves

There are currently 3 candidate sites for the first NPP at the Muria facility on Java Island. The road map for the first NPP includes a public awareness campaign, site permits, regulation and licensing, and ownership decisions, to be completed by 2007. The next step will be to issue tender. Construction is planned for 2010-2015, and commissioning and operation are set for 2015-2016.

Indonesia (BATAN) has a fuel fabrication facility for the research reactor (the Fuel Element Production Installation or FEPI). It is designed to produce fuel elements using imported enriched uranium for research reactors and has operated without significant defect or failure for 20 years. Since May 1996, BATAN transferred all assets of the plant to the state owned company, PT (Batan Teknologi). They plan to develop an experimental fuel fabrication facility. The Experimental Fuel Element Installation (EFEI) will be used for manufacturing and quality control of fuel bundles.

Indonesia is currently analyzing the following initiatives to access nuclear fuel cycle services:

- MNA: Multilateral Approach to the Nuclear Fuel Cycle Initiative of the Director General of the IAEA in 2003, and reported by the Expert Group in February 2005.

- Developing Global Nuclear Infrastructure: Initiative of the President of the Russian Federation, January 25, 2006.
- GNEP (Global Nuclear Energy Partnership): Initiative of the U.S. President, February 2006.
- RANF: (Concept for a Multilateral Mechanism for Reliable Access to Nuclear Fuel), initiative of France, Germany, the Netherlands, Russia, the United Kingdom, and the United States, May, 2006.
- NTI: Initiative of a non-governmental organization, Nuclear Threat Initiative, of Washington D.C., September 2006.

There are six principle criteria for nuclear fuel cycle services from an Indonesian perspective:

1. Correspondence with the Preamble of the Indonesian Constitution: Indonesia will actively promote peace
2. Harmony with the IAEA system: multilateral, integrated safeguard system consisting of a comprehensive safeguards agreement and the Additional Protocol
3. No contradictions with the NPT
4. Non-discriminatory
5. Indonesia supports the right of the Parties to the NPT to undertake research and development (R&D) for peaceful purposes and to fulfill IAEA integrated safeguards agreements and the NPT
6. International assurance for developing countries (party to the NPT) for long-term, timely access to nuclear fuel cycle services (uranium enrichment and reprocessing), for their NPPs

There are several constraints on nuclear fuel cycle services. Fuel cycle services should be easily provided by the market and domestic capabilities for certain countries. If there are no long-term assurances, there will be a need for domestic fuel cycle facilities. The establishment of domestic services would only be justified (technically, economically) if there is a large NPP capacity. The optimal option beyond domestic services may include bilateral agreements between supplier and user countries (perhaps under a memorandum of understanding) or multilateral nuclear fuel cycle facilities in the region.

Indonesia's NPP milestone is to build and operate 4 NPPs by 2025.

Indonesia is considering the following policy options for nuclear fuel cycle services:

1. Natural uranium – purchase from diversified producer countries or produce domestically
2. Uranium processing and conversion – purchase from diversified producer countries or produce domestically
3. Uranium enrichment – purchase from diversified producer countries
4. Fuel fabrication – first loading from diversified producer countries, long-term leasing, or domestic production if economically viable
5. Spent fuel storage – store at plant in the short term and at a centralized facility in the medium term
6. Radioactive waste – processed and managed at the plant or a centralized facility

7. Indonesia will adopt a once-through fuel cycle policy

In conclusion, nuclear fuel cycle services could generate common industries. The possibility of mutual multilateral cooperation for front and back end facilities in the east Asia region (involving the IAEA and international community) would be an attractive initiative.

Discussion

Ivanov: Russia is developing a floating NPP. Would this be useful for Indonesia? Russia would supply the plant and 40 years of support (including take back of spent fuel). Have you considered this?

Karyono: Indonesia's regulatory body (BAPETEN) requires all NPPs to be on land. Also, we would only consider proven technology for electricity (at least 3 years demonstrated proof). What would the level of uranium fuel enrichment be for floating reactors?

Ivanov: 17% enrichment at the lowest.

Karyono: A small NPP is interesting for electricity generation in the remote islands of Indonesia. However, before using a floating NPP, Indonesia should consider if it is also in harmony with the Treaty on the South East Asia Nuclear Weapons Free Zone, ratified in 1997.

Rauf: You have said you will rely on the world market, though fuel fabrication may be done domestically. If supply is denied for political reasons, what strategy would you use to deal with it?

Karyono: Up to this point we have had only a small nuclear enterprise with LEU for a research reactor. We have contracted out for enrichment services, and so far we have not had any problems.

Forrstrom: What happens after interim storage of spent fuel?

Karyono: We have a once through fuel cycle policy with interim storage first. We have many uninhabited islands.

Prepared Remarks of Mohamed Shaker (Egypt)

The IAEA Expert Group studied the question of assurances of supply, which twenty years ago, in 1987, was being discussed in the IAEA Committee on Assurances of Supply (CAS), which went into abeyance that year. CAS was unable to reach consensus on both the "principles for the international nuclear energy cooperation and nuclear non-proliferation" and on "emergency and back-up mechanisms." The 1987 United Nation's Conference for the Promotion of International Cooperation in the Peaceful Uses of Nuclear Energy also failed to reach agreement on such a set of principles. As president of the conference, I tried hard to achieve consensus on such principles, to no avail. I believe we have a good chance this time, after 20 years, to move forward and tackle this issue in a constructive and creative way. We have a lot of food for thought in the valuable report of the IAEA Expert Group on multilateral approaches to the nuclear fuel cycle. We also have a number of initiatives and proposals put forward by the Director General of the IAEA and a number of leaders, which I hope we have a chance to examine more thoroughly during this workshop.

I will begin with a few words on the motivations and prospects for reviving the nuclear power program in Egypt. Twenty years ago, Egypt was about to make its choice of its first nuclear power plant, but failed to do so in the aftermath of the Chernobyl accident in the Ukraine in 1986. After Egypt's ratification of the NPT in 1981, it negotiated a number of cooperative agreements with leading supplier states to begin the implementation of an ambitious nuclear power program. Most of these cooperative agreements are of long duration and are still valid. After a long lull, which went beyond 20 years, the nuclear power project is being reconsidered in the context of the energy mix in Egypt for reasons and factors similar to those existing in other countries. Also, one cannot miss today's renaissance in nuclear energy. There are 29 reactors in 12 developing and developed States being constructed in addition to four units being planned in China alone. This is a great leap forward, which I believe will attract others to do likewise, if their energy needs require such an endeavor.

Egypt decided in 1980 to invest in nuclear power before its great discoveries of gas post-Chernobyl, which brought great relief to the energy sector and more particularly to its electricity needs. It was responsible for the uplift of Egypt's industries and other domestic needs. This was also one reason for the country not to hasten to rekindle its interest in nuclear power. Nowadays, the generation of electricity is mainly dependent on the use of natural gas and oil. In the year 2005-2006, Egypt consumed 17.3 million tons of oil and 541 billion cubic feet of natural gas. Only 12% of electricity is generated by hydro power. Wind energy generates only 1% of electric power. Currently, wind power is of the capacity of 230 megawatts. Next year it is expected to reach 430 megawatts. In 2010, it is expected to generate 3% percent of the total electric power. Egypt is about to establish its first solar energy plant of 150 megawatts.

If Egypt were to invest in a nuclear power plant of a capacity of 1000 megawatts, this would save us 1.78 million tons of oil or 69.9 billion cubic feet of natural gas per year. In a period of 60 years, which is the average life span of a nuclear power plant, the savings in oil would reach 106 million tons of oil, or 4.2 trillion cubic feet of natural gas. This would also spare Egypt the equivalent of 210 million tons of carbon dioxide. It is noteworthy that the reserves in oil and gas are expected to be exhausted in 15 and 34 years respectively. New discoveries in both sources of energy could extend the duration for a few extra years. The average energy demand for electricity in the last ten years was 7% annually. Last year's demand increased by 10.2%. During 2006, the total demand of electric power was 18,160 megawatts, out of the total capacity of 21,300 megawatts. These figures should indicate the type of studies and comparative analysis that are still being undertaken to determine whether it is justifiable to add nuclear power to the energy mix.

Both the Higher Council on Energy and the ruling National Democratic Party are in the midst of assessing and examining the nuclear power potential in Egypt. No final decisions have been made but the Arab Summit in Riyadh last March recommended that members of the Arab League should coordinate and exchange views on Arab cooperation in the peaceful uses of nuclear energy.

The prospects for reviving the nuclear power program in Egypt are not yet very clear, but are still being pondered. If a decision is made to go ahead with nuclear power, it will be to face their future electric needs in light of the short life span of Egypt's oil and gas resources, as well as the limitations on hydro power; only in cooperation with African neighbors on the River Nile, can Egypt double its hydro power sources.

With regard to the second question concerning what type of arrangements for nuclear fuel provision would be particularly attractive or unattractive in light of Egypt's national interests and concern for the nonproliferation regime, the question reminds one of Egypt's dormant agreement of cooperation with the United States signed in 1982, after ratification of the NPT. According to the agreement, the United States is to provide Egypt with fuel along with a reactor, provided Egypt would return the spent fuel to the United States and compensate them for it. Egypt then had no problem with such an arrangement, which obviously reflected proliferation concerns.

Today, however, the guarantee of a fuel supply as a back up measure is a basic requirement, especially in case of interruptions for political reasons. Also, in some cases it will be better to separate between the supply and building of a nuclear power plant, and the fuel needed for it. The latter could be guaranteed through reliable arrangements with the IAEA or regional organizations that would guarantee the fuel supply to its participants in the fuel cycle.

This leads to the third question concerning the different proposals and initiatives, whether by the Director General of IAEA or by a number of leaders. I argue that we ought to decide whether the material to be assured or the material to be guaranteed is the nuclear fuel itself or the enriched uranium, or both. Most of the initiatives and proposals are concerned with the supply mechanism. None has dwelt thoroughly on the merits of a multinational or regional nuclear fuel cycle as suggested by the Director General of the IAEA in 2003.

I fear that one or two of the initiatives or proposals may accentuate the divide between the haves and the have-nots. It is very important to guarantee that any assurance mechanism would not result in a real or perceived division between those two categories of States.

- In any future mechanism there should be a role for the recipients of technology and materials together with the suppliers.
- Any role to be played by the nuclear-weapon States as guarantors of supply would be more effective and credible if these States would also take steps towards nuclear disarmament.
- Article IV of the NPT, and especially the inalienable right enshrined in it for peaceful uses should be re-emphasized clearly and categorically. There are interpretations and even attempts aiming at diluting the provisions of Article IV. They run counter to the spirit of finding ways and means to guarantee the supply of nuclear material and equipment to all those who abide by the nuclear nonproliferation regime.

What is the future of the Nuclear Suppliers Groups (NSG) in this context or in the new arrangement that would ensue as a result of these proposals and initiatives? On many occasions, I have emphasized the importance of a dialogue between the NSG and the potential recipient States before new guidelines have been set forth by the NSG. I believe we need a fair mechanism, hopefully a transitional one towards the new arrangement.

In conclusion, I offer these remarks in all frankness and sincerity. I believe that we are embarking on a very important phase that ought not be wasted and disrupted like other ventures in the past.

Discussion

Walker: At the time it was signed, the 1982 agreement between Egypt and the United States satisfied Egypt's interests, but today a guaranteed supply is a basic requirement. What has changed?

Shaker: We still have a U.S. agreement in force (it is 30 years in duration); and if we were to revive the program and if the U.S. were the first partner, Egypt would not request a change in the agreement unless the United States and Egypt agree that they should reexamine it. In any of our agreements (e.g. with Australia, Canada, France, Germany), it is important to have solid guarantees of supply of fuel. This is a fast-changing world and relationships change quickly. We should have a backup, automatic mechanism in case fuel or equipment supply is disrupted.

Rauf: You mentioned an amendment to Article IV. Could you explain what you meant by this?

Shaker: Without Article IV, there would be no NPT. Article IV is more important than Article VI. Article IV gives the right to pursue uranium enrichment. For example, Iran has this right, but could defer the right. If Article IV were to be amended it would be for the sole purpose of strengthening the inalienable right rather than weakening it by devious interpretations.

Bunn: Could you outline the things that have changed and lead to your optimism? If a regional center were established in the Middle East, would participation by Iran be acceptable?

Shaker: I am not "optimistic," but 20 years ago the focus was only on the principles of cooperation. Today we have bold proposals from the Director General and other leaders. These did not exist in the 1980's. There has also been some other encouraging work. The question now is whether or not the political will is there (and not just technical know-how). With regard to Iran, for years Egypt has been a proponent of a Mideast weapons-of-mass-destruction-free zone. This would extend to all Arab countries, Israel, and Iran. At the Arab summit, we looked for a mechanism to make this 20-year dream a reality. There is now determination to find such a mechanism, which could be revived by reengagement for the settlement of the Palestinian-Israeli conflict. Iran could be brought in.

"The Energy Sector of Armenia," Areg Galstyan (Armenia)

Loss of energy security is a subject of great risk for Armenia, which is situated in a difficult geopoilitical zone but is keeping its political and economic stability. The impact of energy security loss on the social-economic life of Armenia can be assessed by the bitter experience gained during the energy crisis of 1993-1995.

Armenia is wholly dependent on outside energy sources. The only domestically produced primary energy is electricity from hydroelectric plants and, conditionally the single nuclear plant (nearly 45%). The closure of this NPP in 1989, caused an energy crisis, leading to an increase in demand for hydropower and subsequent ecological harm. Therefore, the second unit (of two) at the NPP was restarted in 1995, allowing Armenia to overcome the crisis.

In 2005, a new strategy for the period to 2025 was announced, calling for nuclear and renewable energy. The strategy aims to achieve sustainable economic development in Armenia;

enhance the energy independence and security of the country including diversification of imported and domestic energy resources; and ensure efficient use of domestic energy resources and the development of renewable energy sources and energy savings.

The "Least Cost Generation Plan for 2006" (LCGP) was developed in 2006, with the assistance of the U.S. Agency for International Development (USAID) based on the principles of the "Economic Development of the Republic of Armenia within the Framework of the Energy Sector Development Strategy," which was approved by the Armenian Government. After considering a number of development scenarios, incorporating gas and oil price changes and environmental impact, they concluded that nuclear energy is the only option for base-load capacity in Armenia.

Guided by the implemented analysis, as well as by strategic and economic research, the following recommendations are made in the LCGP:

1. decommission the Armenian Nuclear Power Plant (ANPP) in 2016 or earlier, as soon as the new nuclear energy unit is ready
2. complete funding of ANPP safety upgrade projects and the required investments to ensure safe operation of the nuclear plant before its decommissioning
3. complete a comprehensive safety and environmental assessment of the ANPP site to determine compliance of the site with decommissioning and construction requirements for the new units
4. develop a comprehensive decommissioning plan that shall be implemented five years before the commencement of ANPP decommissioning and shall be based on the provisions of the ANPP Decommissioning Strategy approved by the Armenian Government
5. determine funding sources for ANPP decommissioning, form a decommissioning fund, select the fund's manager who will manage it until control over the low risk investments is switched to international organizations
6. develop and implement a plan targeted to resolve problems regarding Armenia's ability to finance and construct a new nuclear plant by including size and allocation issues
7. develop local renewable resources to enhance energy independence and ensure diversity of energy sources
8. develop and implement projects that encourage energy efficiency, making this sector attractive for consumers and contributing to the acquisition of energy efficient equipment and devices
9. establish and implement a project to minimize the impact of tariffs on consumers with regard to the commencement of ANPP decommissioning and new nuclear capacity

Any decision about the future status of the Medzamor NPP should consider the following: the government of Armenia confirms its consistent position on NPP decommissioning and at the time of decommissioning, at least 2.5 billion kWh electrical energy generation should be guaranteed by new capacities for covering electricity demand in base-load, which will grow to 5 billion kWh in 10-15 years. Energy generation should provide for social-economic development and long-term energy demand growth, and any abrupt jump in tariffs should be avoided.

The 2025 strategy calls for a new NPP at the Medzamor site in the near-term. The only way to ensure energy security is to build new NPPs in the republic. Armenia intends to continue investment with its partners. Currently, the NPP produces 2.5 billion kWh of electricity. By 2025, up to 60% of total domestic consumption will be produced by nuclear power. The remaining demand will be covered by renewable (mainly hydro) resources (30% of total demand) and the rest will be covered by thermal resources. To implement new NPP construction, the law can now cancel the state monopoly on ownership of new units (leading to more flexibility and more attractive investment). Specific recommendations include building new units on the same site, continuing investments in maintaining the safety of existing units, and establishing a decommissioning foundation. Funds will come from tariffs on energy from these units.

Fresh fuel is supplied by Russia. The development of a wet storage facility is in progress (with help from Areva). The third stage is being developed to store spent fuel at the NPP site. We do not intend to develop enrichment capabilities, so Armenia is very interested in nuclear fuel cycle services, and it welcomes the formation of international fuel centers under the aegis of the IAEA. As a result of the studies conducted by the Ministry of Energy, the following action plan for the nuclear energy sector was accepted: implementation of all necessary steps toward continuous enhancement of safety levels at ANPP until its decommissioning; preparation and implementation of ANPP decommissioning procedures; resolution of the issue of construction of a new nuclear unit to replace the operating unit at the ANPP.

Through the initiative of the Armenian Government, amendments were made in the Law of Energy and adopted by the National Assembly of Armenia, abolishing the state monopoly. This will allow investments in the construction of new nuclear units from other financial sources too. In our opinion, the role of new nuclear units for base-load electricity as well as the electricity supply to regional countries also provides an opportunity for special financing.

Armenia received an official proposal from the Russian Federation to join the pilot project at the International Uranium Enrichment Center (IUEC) at Angarsk. The establishment of the Center was considered by governments of interested countries on the basis of: intergovernmental agreements with respect to the inalienable rights of countries to the peaceful use of nuclear energy without discrimination; the absolute and reliable adherence to nuclear nonproliferation requirements; and mutual benefits and market relations. Armenia declared its principal commitment with regard to the proposal. However, Armenia's participation largely depends on its Concept of nuclear energy development as well as on the proposed structure and operational functions of the Center.

Discussion

Rauf: Has Armenia looked into long-term fuel supply?

Galstyan: Fuel for the existing NPP will come from Russia until it is decommissioned (under a long-term agreement). For future units, we are only at the feasibility stage. It is too early to discuss this, though the answer may be similar to long-term commitments based on previous experience. We have great interest in long-term supply commitments.

Budnitz: Assuming good relations for the next 10 years, could Armenia be brought into the Russia-Kazakhstan Center in Angarsk? This seems logical but not necessary.

Galstyan: Of course we are aware of the preliminary Russian-Kazakh discussions. Armenia is not yet involved. There is no talk of a definitive agreement, and Armenia has not yet been invited to join.

Myasoedov: Indeed, Russia and Kazakhstan are in full agreement. Russian President Vladimir V. Putin will soon sign an agreement in Kazakhstan. A special mechanism is to be developed to bring in other international partners. This requires a government-government agreement. We already have an existing facility for a center, and there is no need for the development of new capacities. We have yet to complete legal formalities, but participating commercial organizations will share risks and benefits.

Solonin: In terms of fuel supply, TVEL provides a significant portion of the world demand for nuclear fuel. The market is stable. Global producers treat deals very carefully. We can guarantee Russian supply of nuclear fuel over the entire cycle (including uranium extraction).

Budnitz: There is a difference between the current arrangement and a new arrangement, as Russia will take back and keep spent nuclear fuel. This is a positive option for Armenia.

Solonin: We are ready to take back VVER400 fuel and reprocess it at RT-1. Uranium produced is to be used for atomic energy. We must think carefully about transportation because of the geographical location of Russia and Armenia. We are ready to start negotiations.

Ivanov: I have a comment on changes to Russian law. The legislation now allows legal entities to be owners of fissile material. It can be legally transported without a change in ownership. Participating countries can be assured that fissile material ownership will not be changed. Also, Russia can now accept foreign spent fuel for long-term storage. Interim storage is not as good; long-term storage must be faced eventually. Recent legislation is only for Russia; there is no international law governing transport routes. For example, Bulgaria was ready to transport through Romania and Ukraine, but there were problems. Rail transport will be important, as will air (using special containers). The law was changed by the Russian parliament in light of international fuel centers. We must start this work now.

Forrstrom: Dr. Galstyan, you seemed positive about international assurances. Is this true in general, or in the case of political cut-off?

Galstyan: In principle, we are positive in general. Will there be political limitations in the future? We cannot say. Today, we have clear plans about the feasibility of nuclear power in the future. We are not worried about political hindrances.

Gottemoeller: We have heard about the new Russian laws and the new Armenian laws meant to facilitate private investment in NPPs. Will these involve regional cooperation or just focus on private investment?

Galstyan: Last year there were new amendments. Atomic energy will continue to be a state monopoly only for spent nuclear fuel, waste, and fissile material. We are thinking about specific investments. We are trying to attract investment for new units (owned by a state or a private firm).

Rauf: Dr. Ivanov, regarding the Duma amendment and the issue of ownership change: What exactly is allowed? What is new? Will Angarsk provide only enrichment services?

Ivanov: Here is what is new: before, fissile materials on Russian soil could only be owned by the state. Therefore, TVEL-held uranium required government permission. For example, natural uranium, before the new law, had to be state-owned. Therefore, intergovernmental agreements were needed to guarantee return. Now, private bodies can own fissile material in Russia, if they are listed as allowed and experienced to do so (by the Russian President). (There are new uranium mining negotiations underway with Japan.) Regarding Angarsk: this will be an enterprise providing enrichment as specified by the customer, no natural uranium will be dealt with (other plans handle this).

Myasoedov: Regarding Russian-Armenian state cooperation, sometimes, relations between states are not good (e.g., Russia-Georgia). An international center is not a state-state agreement. It is a commercial entity, with an independent council stressing equal rights for all participating countries (perhaps with an IAEA representative). Agreements will be made with the Angarsk enterprise. However, enrichment technology cannot be transferred. This is an unconditional demand.

"Australia's Current and Future Nuclear Fuel Cycle Activities," Ian Smith (Australia)

Australia has more uranium than any other country and is interested in selling it. The assurance proposers need to involve people who have the bulk of the world's uranium. There is increasing political interest in the nuclear industry prompted by climate change, international nonproliferation commitments, uranium prices and Australia's uranium deposits. Australia has much of the world's "low cost" uranium resources and there is high potential for future discoveries. Australia provided more than 20% of world production in 2005, and can increase uranium exports significantly. Australia acknowledges the potential value-added to uranium mining by conversion, enrichment, etc. The recommendation however is to not pursue these options at this time but to retain the possibility for the future. The government is open to supporting an international fuel bank as a means of limiting the spread of proliferation sensitive technologies by providing fuel supply assurances and allowing the expanded use of nuclear power.

There is increasing political interest in the nuclear industry, prompted by climate change, nonproliferation concerns, and the high price of uranium. Recently, a review of "Uranium Mining, Processing, and Nuclear Energy in Australia" was conducted. Though Australia contains the world's largest uranium reserves (low-cost, high quality uranium), it produces less than Canada. Uranium is currently mined at only 3 mines; another mine is approved and will open in 2008. This is due to politics and "artificial blockages." Australian uranium resources include:

1. 12,360 tons U_3O_8 exported in 2005

2. Low cost reserves (94,000 tons at Ranger, 21,000 tons at Beverley, 1.6 million tons at Olympic Dam)
3. Potential resources over 1,400,000 tons

The recent government review found that uranium resources are plentiful (including potential for future discoveries); uranium exports can be increased significantly; and impediments to the expansion of uranium mining are recognized and being addressed. It was also acknowledged that conversion and enrichment could add value to uranium resources (though there would be investment challenges and challenges in accessing enrichment technology). The recommendation is to not pursue these options now but to retain the possibility for the future.

Australia has a very strong commitment to the NPT and safeguards, with the strictest uranium export controls in the world, requiring a bilateral safeguards agreement before supply. Only 11 countries receive Australian uranium, with conditions set through bilateral treaties. (Negotiations are underway with China.)

The government is open to supporting an international fuel bank as a means of limiting the spread of proliferation-sensitive technologies by providing fuel supply assurances and allowing the expanded use of nuclear power. This is a good commercial opportunity. Challenges will include integrating this into the current market, transportation of spent nuclear fuel, retention of the ability to enforce bilateral arrangements, and maintenance of strict export controls. A full analysis of the implications of such a system has yet to be completed.

There is an unprecedented level of public awareness of climate change in Australia, and therefore an understanding of the need for greenhouse gas reduction. This has led to the consideration of nuclear power. Questions remain, however, about how to handle nuclear waste, including the cost. A government review scenario suggested a reduction of 8-17% of CO_2 emissions by 2050 due to the construction of 25 NPPs. Potential investors in the nuclear power industry would require a stable policy environment and a predictable licensing and regulatory scheme.

Australia's nuclear waste disposal policy does not allow acceptance of waste from other countries or the use of reprocessing. Reprocessing is unlikely to be attractive for Australia until warranted by a domestic industry. An advanced ceramic waste form technology (synroc = synthetic rock) has been developed by the Australian Nuclear Science and Technology Organization for long-term immobilization of high level waste (HLW). Synroc is a waste form built on natural minerals that have demonstrated their survival over geological timeframes and are highly-proliferation resistant. There are tailored ceramic and glass-ceramic waste forms for problematic waste streams; this suite of waste forms has become internationally recognised as a de-facto performance baseline for HLW waste forms.

In conclusion, the government review has looked at all aspects of uranium mining and nuclear power in Australia. A cabinet-level response is imminent. Australia is most involved in the front and back ends of the fuel cycle through the supply of uranium and technology respectively.

Discussion

Solonin: Are there plans to set up spent fuel repositories in Australia?

Smith: No.

Bunn: You stated that in terms of adding value to your uranium, the current recommendation is to not pursue this now, but to leave it open for the future. Correct?

Smith: We believe Australia has the potential to be a major supplier.

Lowenthal: Could you explain the uranium export controls you mentioned?

Smith: There are currently rules against reshipment, reprocessing, and about enrichment levels. There is also an independent accounting process. Also, there are fallback provisions if IAEA safeguards break down.

Myasoedov: Will Australia send uranium to enrichment centers?

Smith: There has been no decision yet. It is not clear that there are such strict disposition requirements with the centers.

Gottemoeller: In Moscow, there have been reports of Russian-Australian cooperation on uranium exports. What is the status of these? Are changes in Australian law required to allow for foreign investment?

Smith: Australia wants to keep the option open. It supports the NPT and believes a fuel bank could help this. I still have not seen the details. Australia supports every country's right to develop nuclear power, but none of the proposals has safeguard requirements that meet Australia's standards.

Gottemoeller: Would foreign investments be allowed?

Smith: Yes. Foreign investment is allowed and exists. Recently, a Russian delegation visited Australia, but these types of negotiations take a long time.

McDonald: It seems that none of the current proposals have safeguards requirements strict enough for Australia.

Smith: We welcome this opportunity, and would like to be involved in the discussion.

McDonald: Would Australia object to others pursuing assured fuel supply with other uranium?

Smith: No, but the countries with uranium reserves were not included in the initial conversations, which was a major omission; we want to be part of the conversation now.

Pedro Raul Villagra-Delgado (Argentina)

The importance of nuclear power's potential for sustainable energy development, as well as for areas such as medicine, improving conservation of foods, etc., is being revisited and there is a clear tendency to its increased use in the future. All countries have the right to benefit from its potential uses for exclusively peaceful purposes, in line with their own national priorities and objectives, and in conformity with international law, the general rules on nonproliferation, and the need to eliminate threats to international peace and security.

The NPT created two categories of states. That logic should not be replicated in other agreements. No new system based on discrimination against those who comply with the international rules on nonproliferation has a chance to be considered legitimate. Nuclear technologies are not intrinsically bad. There is no reason why a kind of technological privileged group should be established to the detriment of the rest of the whole international community nor any reason why a country with impeccable credentials like Argentina should renounce its legitimate rights to develop nuclear energy for peaceful purposes if it deems that it may be conducive to its technological development. Any approach that may imply that countries not willing to renounce their rights to develop elements of the nuclear fuel cycle in the future may encounter difficulties in accessing the market for nuclear fuel may act as an incentive to developing such facilities and the very notion of it may be in breach of NPT Article IV, which recognizes the pre-existing inalienable right of states to develop research, and production and other uses of peaceful nuclear technologies. This right is not a result of the NPT but the very notion of statehood. The Argentine document (NPT/CONF.2005/W.33) of the 2005 NPT Review Conference contains very useful ideas on this matter, which I fully share.

The objective of developing either international mechanisms based on facilities owned by any given State or through international centers should be clearly determined, whether it is to curb proliferation or to provide a reliable system to guarantee supply of nuclear materials to allow the development of nuclear energy in the years to come. If the purpose is the former, the proposed mechanisms may prove to be unsuitable as they will be targeted toward those who are in compliance with the international nonproliferation regime with very little or no chance of engaging those countries who do not. If their purpose is to facilitate the development of nuclear energy on a mutually convenient basis, these mechanisms could be of interest for some countries.

For those who intend to violate their NPT obligations, any scheme intended to develop multilateral approaches to the nuclear fuel cycle will most likely be irrelevant as they will most surely not participate in any mechanism which could impinge on their actions intended to acquire nuclear weapons. If their purpose is to facilitate the development of nuclear energy on a mutually convenient base, these mechanisms could be of interest for some countries. For those countries, which represent the vast majority, there is no problem in acquiring nuclear fuel abroad today. The market is functioning well and it would be advisable not to disrupt it.

Urenco explained during the International Group of Experts that they could not provide such guarantees as the final decision on whether or not to export would not depend on the license issued by the country where the exporting plant was located. Neither could they export to a fuel bank, as only final recipients could apply for their licenses. In any event, guarantees of supply should have an international nature and therefore be made by treaties instead of mere commercial contracts, as was the case with Urenco.

Argentina is going to build a third and fourth NPP. There is a debate as to whether restricting enrichment technology is the right approach or if we need improved safeguards.

In 2004-2005, I participated in the IAEA Expert Group on "Multilateral Approaches to the Nuclear Fuel Cycle." Its report of February 2005, is the most thorough study on all aspects of the multinational approaches and should be followed. Paragraphs 312 and 321 shed light on the political aspects of it. We must take into account (1) access to nuclear energy and (2) the nonproliferation regime. I am more optimistic than most. The scope of the IAEA work shows that there are only a limited number of "problem states." They have been the same ones for many years. Only two countries are under IAEA scrutiny on this matter. Any action proposed needs to recognize a preexisting right to develop any non-weapons technology under Article 4, paragraph 1.

The question remains, will the numerous proposals [on assured supply] help the nonproliferation regime? Will a system of guaranteed supply interfere with the existing market even in countries with appropriate safeguards? They can only be established on a voluntary basis. Otherwise, they will be contrary to the NPT. New conditions under which States access nuclear fuel or technologies for peaceful uses should renounce their rights under Article IV of NPT, even if in full compliance with their nonproliferation obligations, may constitute a breach of NPT Article IV, by those demanding such new conditions, particularly by nuclear weapons states who, it could be argued, are themselves in breach of Article VI. A mandatory system could only be based on rules applying to ALL States and ALL facilities, including nuclear weapon states.

Argentina will not resign forever any nuclear fuel cycle rights but will support any safeguards, either current or enhanced. Concerning regional centers: what is their political context? There is concern about locating them in nuclear weapons states. In addition, disarmament must be addressed. Article VI must be complied with, progress on the Comprehensive Test Ban Treaty and the Fissile Material Cut-off Treaty must be made, and the 13 steps agreed on at the NPT 2000 Review Conference most be implemented. Any idea corroding the delicate balance of the NPT may unravel the whole nonproliferation regime.

Argentina does not see the need for multilateral approaches to the nuclear fuel cycle because there is no problem with markets. In the early 1990's, much forward progress was made. The Additional Protocol was developed cooperatively after the first Iraq war, but stalled at the end of the decade.

Regarding "breakout" (withdrawal from NPT with 3 months notice) the system is actually not as flawed as portrayed. It is clear that Article XI of the NPT brings the matter of withdrawal from the NPT to the United Nations (UN) Security Council and this body should act if it deems that a country being party to the treaty developed nuclear weapons, as that would be a breach and a conceivable a threat to international peace and security. The problem may be that the U.N. Security Council does not act when it should, but then no multilateral nuclear approach is going to solve this. Multilateral nuclear approaches trying to stop proliferation will not work. They may, on the contrary, help increase access to nuclear energy.

Discussion

Gottemoeller: What is your view of what should be done to get Article VI back on track?

Villagra-Delgado: We can't do much without the commitment of the main actors to eliminate their nuclear weapons. There was much progress in the 1990s, but they are now stalled and even losing ground with talk of improving armaments. But without progress on Article VI, there can be little progress on other issues. It is an obligation under the NPT and it has to be seen as a mandatory requirement.

"Brazilian Utility Perspective," Leonam dos Santos Guimarães (Brazil)

Brazil is famous for many things including beaches and carnival. There is another side of Brazil that you must know: innovation, technology, competitiveness, and productivity. This includes the "nuclear Brazilian industry." Brazil has a synergetic mix of large uranium reserves, fuel cycle technology, pressurized water reactor (PWR) technology and nonproliferation. This could provide an important contribution to "assurances of supply," both globally and locally.

Nuclear power in Brazil is a state monopoly established by the constitution. The nuclear fuel industry in Brazil has uranium and technological capabilities: mining and milling; conversion; enrichment; re-conversion; pellets; fuel fabrication; NPPs. Uranium is considered a public asset. Today, Brazil operates two NPPs (Angra-1 and Angra-2) and Brazil will commission a third NPP (Angra-3). In the future, they will be a leader in R&D, and will operate 4 NPP (with 2 turbo generators). There is an unprecedented level of public awareness of climate change impact due to intense drought in much of Brazil and the fact that there are water shortages and restrictions in most major cities.

Brazil is a big country (in terms of area, population, and gross domestic product) and the 10th largest electric generator (and 2nd largest hydropower producer) in the world. However, it is only the 90th largest energy consumer (per inhabitant). Brazil relies a great deal on hydropower, but this needs "thermal regulation" (to compensate for seasonal fluctuations in water stock/storage). This drought was the root cause of the 2001 electricity supply crisis. This tendency will be amplified by expansion of hydropower in the Amazon basin. Brazil has access to thermal fuels, including coal, biomass, natural gas, crude oil, and uranium. Costs, volatility, and assurance of supply are factors to consider. Brazil needs more thermal generation to stabilize the system.

Brazil is one of few countries with large uranium reserves and local production, and has full, open fuel cycle technologies. Brazil also has nuclear power technology and is concerned about nonproliferation. Only 30% of our territory has been prospected. In the end, we expect to be among the three leaders in uranium production. The Lagoa Real mine assures supply for the Angra NPPs and any new ones through 2030. The Itataia mine should be developed for international markets.

Currently, Brazil is operating and maintaining the Angra 1 and 2 NPPs, which produce 657 and 1,350 MW of power respectively. In the near term, engineering, procurement, construction and commissioning are expected to begin for the Angra 3 NPP, which is to produce 1,350 MW. Further R&D for nuclear power will continue.

Brazil's medium-term vision includes investing profits in industrial development, aiming to achieve self-sufficiency and "added value" to exports; no planned reprocessing (abandoned 30 years ago); and development of long-term interim storage of fuel assemblies (deferred decision on fuel cycle). Brazil has two demonstration enrichment plants and one user plant under construction. The long-term vision includes: continental integration and assuring regional supply of uranium and (open cycle) nuclear fuel services with full scope safeguards. Brazil's decisions about expanding or ceasing enrichment and fuel fabrication are not based on the profitability of the enterprise; price volatility and assurance of supply must be considered as well. Conversion, enrichment, and fuel fabrication amount to about 30 % of the fuel cycle cost.

The Brazilian constitution calls for peaceful uses of nuclear energy. We have a remarkable record of more than 25 years without technical deviations or suspicious events. Like Germany, Japan, and the Netherlands, Brazil has "2+1" enrichment plants fully safeguarded. We ask, how much additional separative work unit (SWU) is needed for the nuclear renaissance? Do current facilities have enough capacity?

In conclusion:

1. Any solution envisaging limiting the access of some countries to technology will mean assuming the bankruptcy of the international nonproliferation regime.
2. The IAEA is not a commercial enterprise, limiting chances of success.
3. Brazil has large uranium reserves, technology, and fully safeguarded industrial facilities for all open fuel cycle steps and could play an important role in future IAEA assurance of supply mechanisms such as a hosting a regional production center.
4. The problem is not assuring supply but assuring political stability. Only democracy and development can do this.

Discussion

Kelly: What is "long-term interim storage?"

dos Santos Guimarães: We adopted long-term interim storage to gain public acceptance. We have a long-term 500 year storage strategy. We have no intention of reprocessing.

Kelly: Are there international or regional elements to your back-end strategy?

dos Santos Guimarães: It is international only in the establishment of a regional enrichment center. The spent nuclear fuel storage site is just for Brazil.

Ivanov: Argentina's and Brazil's positions are understandable: defending their countries' interests. This is not on the agenda of internationalization. For example, take the policy of delayed decision. Russia's strategy is to set up a temporary site for irradiated fuels to be used later in fast breeders – a valuable product. Internationalization is a concept under which irradiated uranium may be stored. If interested, Russia would offer its services to others. Russia stands ready to offer solutions if interested. Long-term shipment is not an option.

Ahearne: Ambassador Shaker [from Egypt] mentioned regional cooperation in the Middle East. Has there been any such cooperation in South America?

dos Santos Guimarães: Argentina and Brazil were pioneers in cooperation. We developed nuclear energy by ourselves, first without safeguards, but have had a common nuclear policy since the late 1980's (for example, accepting foreign waste not acceptable). Bilateral cooperation will continue without the label of "regional cooperation."

Villagra-Delgado: Brazil and Argentina are pioneers in regional cooperation and offer a good model. There is a bilateral system of safeguards and common nuclear policy, both domestically originated. It is difficult to imagine better cooperation. In terms of waste, we are constitutionally prohibited from receiving foreign waste. It is tough to secure public acceptance of even domestic waste. The Russian proposal is interesting, but it does not improve nonproliferation. The IAEA safeguards system works very well, with only a few exceptions that have been clearly detected. Regional cooperation and commerce with IAEA safeguards are the best way forward.

Ahearne: What about Chile, Peru, or Uruguay (i.e., multilateral cooperation)?

dos Santos Guimarães: Peru is the only other South American country with nuclear technology (for medicines). The other countries are too early in their development now. Cooperation will come naturally. Also, I must mention the Tlatelolco Treaty, establishing Latin America as a nuclear weapons free zone.

Villagra-Delgado: Chile has been considering nuclear energy and its first NPP. This could be a good opportunity for multilateral cooperation. They should approach Brazil and Argentina to learn from our experience. Malaysia also sent experts to Argentina to learn about "reactor operators." This is another good opportunity for "multilateral" cooperation. ABACC is always ready to help.

Shaker: You are looking into the future. Urenco and Eurodif are good [multinational] examples. Brazil and Argentina could operate something like this. Can Latin America, or Japan (and the Far East) set up new regional centers for uranium supply as well?

Solonin: Let's assume there exists an international nuclear fuel cycle center, which can guarantee fuel over a reactor's life cycle and which is economically favorable. Would Argentina and Brazil use the center or their own facility?

Villagra-Delgado: Yes, this is potentially very attractive, but this is not just an economic decision. All technological developments have spin-off potential for the country in other areas. Can long-term contracts be provided if no facilities are developed? We are not sure, probably not. Will we renounce our rights to technological development? This is out of the question. We do not want to go backwards and watch our capabilities die. We will not settle for sub-standard teaching at our universities.

dos Santos Guimarães: Thirty years ago, the same deal was made with the U.S. This was broken in the 1980's. This is a fact. This offer now is too late. Brazil is already constructing an enrichment plant. I am sure that the domestic costs are lower than imports from Russia.

"Nuclear Story of Korea," Chang-Kook Yang (South Korea)

Nuclear power in Korea is motivated by energy security, diversification of energy sources (from 1973 to 2006, foreign energy dependence rose from 56% to 97%—during this time, nuclear energy rose from 0% to 39.4% and gross national product increased 45 times), economics, national pride, reduced oil dependency after the 1973 oil crisis, and reduced greenhouse gas emissions.

Nuclear power in Korea began in the 1970's with turnkey contracts with foreign vendors. In the 1980's, they slowly learned this technology from vendors with component-based contracts. In the 1990's, joint development of new fuel types was pursued with foreign vendors promoting export of fuel assemblies. Also in the 1990's, Korea achieved 95% self reliance of design and construction technology for NPPs. In the 2000's, they are promoting the export of nuclear power plants and today they operate 20 NPPs, with 6 under construction (4 OPR1000s and 2 APR1400s) and 4 more are planned by 2020. Korea does not have a program for localization of the CANDU technology; the OPR1000 and the APR 1400 have been developed by local entities.

In terms of the nuclear fuel cycle, all uranium concentrate is imported. Since the 1950s Korea has looked for uranium but has failed to find economically recoverable resources; 100% of resources have been imported. To enhance security, a diversification of supply sources has been pursued. Reliability of supply is one of the most important considerations, so they would like to enter into long-term contracts with reliable suppliers. They have no plans for domestic enrichment services, and therefore have contracts with Urenco, Eurodif, and TENEX. Korea has developed the next generation of fuel for the Westinghouse type reactor, OPR 1000, and APR 1400 with foreign vendors. All fabrication services for Korean reactors can now be provided by local entities.

Enrichment and spent fuel reprocessing is divided between "haves" (pursuing technology transfer on commercial basis among them) and "have-nots" (with strict restrictions on such cooperation). Internationalization of the nuclear fuel cycle should be based on industry experience. From a commercial point of view, nuclear fuel supplies have evolved to provide better services. Fabrication technology can be acquired on a commercial basis or through joint R&D. South Korea has never experienced fuel supply cut-offs. In the existing market, suppliers would respect a company's reputation. Even if supply is disrupted, there are alternative sources. The existing market for fuel has functioned so well even though supply disruption occurred for several reasons, utilities purchased nuclear material from alternative sources. Nuclear power plants could be operated without a stop.

The concept of a nuclear fuel bank is acceptable.

They see three prerequisites for an assurance of supply mechanism: immediate supply of nuclear material in the case of supply disruption without political consideration of NPT compliant nations, a fair market price, and not hindering the existing commercial market. Providing reassurance to fuel supply will reduce the risk of proliferation.

The following are problems to be solved before assurance of supply proposals can be realized: international consensus must be reached, financing must be determined, technical problems (proliferation-resistance reprocessing, burner reactors, etc.) must be addressed, and industry must participate.

In conclusion, the commercial market has answered industrial needs well and such confidence will continue in the future. However, the proposals raised by several countries will give additional assurance of supply to newcomers (as a last resort). That will reduce the temptation of conducting R&D in sensitive technologies for additional security of supply and reduce the risk of proliferation.

Discussion

Fetter: What is the enrichment policy in South Korea? Reprocessing policy?

Yang: Korea currently has 16 PWRs and is planning 6 more. Korea needs 2M SWU/year, but has no plans to construct enrichment facilities now. In terms of reprocessing, we maintain a wait and see policy, and spent fuel is stored at onsite interim storage facilities. It took 20 years just to locate a low-level waste disposal site. We are in the early stages of R&D for pyroprocessing, which is different from wet technology (more proliferation resistant).

Gottemoeller: Has your emphasis on joint R&D work for "next generation fuels" added to interest in cooperating with international centers if there are enhanced opportunities for collaboration (e.g., on next generation technologies)?

Yang: As the fabrication market is under severe competition, Korea now needs more economic and safer fuel, but the development of new fuel is too expensive for one company. We have no plans to develop enrichment or wet reprocessing technology.

Villagra-Delgado: This is an important point. There is good cooperation between many countries, and it could be an enticement. But some proposals are aimed at preventing the development of technology, which is wrong. Knowledge is not inherently bad. Training of people is important. Does it outweigh the proliferation risk? I am not sure. Argentina is not interested in a turnkey system.

dos Santos Guimarães: There has been multinational cooperation on fuel development. Opportunity is a market question.

Gottemoeller: What about new technologies not yet on the market (such as Generation IV)?

dos Santos Guimarães: Generation IV is a good example. There is still a long way to go before industrial use.

Smith: In Australia, R&D of advanced technologies is important to enable us to be an "informed procurer." For example, we recently purchased a research reactor from Argentina, and were able to buy a world class reactor at an attractive price.

Budnitz: In regard to "confidence in the common market," how do your NPPs behave in terms of ordering in advance? How much fuel is ordered, and how far in advance?

Yang: Deliveries are received 4 months before reloading. The lead time is based on each individual contract, though we plan to maintain approximately a 1 year supply in the inventory for 20 NPPs. We believe this will help to manage a disruption.

Budnitz: We have just heard that a customer desires 1 year of assurance, but is confident that the market will be reliable.

Bunn: But Brazil could say that the United States cut off supply to Brazil, and it took many years to overcome. Under these circumstances, you might need more supply on hand until you can find another supplier.

Budnitz: This was a disruption when Brazil had only one plant.

dos Santos Guimarães: Angra-1 is a turnkey plant from Westinghouse with a government-to-government agreement, and fuel is assured by the contract. When Brazil changed its policy from turnkey to technology transfer, the U.S. government used assurance of supply to put pressure on Brazil. Brazil turned to Germany to develop fuel, and CAVEU had to produce fuel for a Westinghouse reactor, which was difficult. Brazil mines enough yellowcake for Angra-1, 2, and 3. This is then sent to Canada then to Europe then back to Brazil. We say that "uranium needs an I.D. AND a passport!" This chain could be disrupted for many reasons. We do not have a big inventory.

Budnitz: We are talking about a political interruption of Angra-1 [Brazil], but what would happen to a country like South Korea, which gets 40% of its electricity from nuclear power? This would cripple the country. Yet they have confidence in the world market.

Bunn: Why don't you keep a larger inventory given your experience with the United States?

dos Santos Guimarães: Brazil lies outside of today's "political hotspots." We fulfill all of our nonproliferation requirements, and are not concerned about politically-driven cut-off. Additionally, it is too cost prohibitive to have a larger inventory.

Villagra-Delgado: The market works very well (aside from a few bad apples). We must try to prevent an assurance mechanism from becoming a hindrance to the market: "If it isn't broken, don't fix it."

Myasoedov: This is a good discussion with countries having ambitious plans. We see more and more irradiated fuel, which will give rise to new problems, such as terrorism and radiation dangers (especially in densely populated countries). We cannot convert spent fuel to fuel elements. There should be a strategy for spent fuel.

Yang: Spent fuel is a headache; even an interim storage site is difficult to locate. Korea has no plans to reprocess spent fuel. For an international center of this type, what would they charge? Without this information, I cannot answer the question.

Myasoedov: Russia is now searching for approximate answers. We cannot postpone answers forever.

Shalabi: Fuel supply arrangements are negotiated country-to-country. What about the licensing process and regulatory requirements?

Villagra-Delgado: As far as supply of fuel is concerned, for a receiving country to be assured of quality, the licensing country must be guaranteed that the licensed company has good quality control. This is a well-tested practice, following the example of other international companies. This process makes it possible to see if fuel meets requirements and standards.

Ahearne: Should we be optimistic or pessimistic or cautious? Will the Russian program handle all problems? What about the Direct General's fuel bank proposal?

No response.

Levenson: I would like to hear comments from speakers about how far in the future they might consider fast reactors.

No response.

Ivanov: Russia has operated fast reactors for over 30 years, and it is a good example of how you can deal with nuclear waste and close the fuel cycle. At this point we have problems with fast breeders. The BN600 in operation meets safety requirements. We've tested different types of fuels, looked into the volume of nuclear waste, and closing the fuel cycle. What about the BN800? Russia expedited funding, and plutonium was produced and accumulated. The nuclear weapons states must lead the example in disarmament. Our attitude should be determined for each particular country. Russia welcomes cooperation on the future of the nuclear fuel cycle in relation to breeder reactors.

Galstyan: In Armenia, we are aware of the fact that we are consumers of nuclear power. These proposals and international centers are a promising start to addressing the problem of handling irradiated fuel. We would welcome new emerging tendencies.

Shaker: A back-up mechanism should not be limited to political problems. There exist other types of disruptions. I think the mechanism should be triggered for other reasons. We should look at a bank in terms of a larger framework. Will it be an IAEA bank? Controlled by the Board of Governors?

Villagra-Delgado: All these questions were discussed at length and in depth in the IAEA Expert Group Report issued in February 2005. We should pay more attention to it. Any fuel bank must be a *virtual* bank, since there are too many types of fuel. Every reactor needs its own. It is technically incorrect to say that we can have ONE fuel bank for all. Any bank must be based on treaties, not contracts between states (and the IAEA). It should not be based on the whim of any country.

dos Santos Guimarães: I have the impression that we have a solution to a poorly defined problem. The threat is not clear. We must define the problem. No historic proliferation threat was due to assurance of supply. What exactly is the problem?

Solonin: Consider the IAEA INPRO project, initiated by Russia. This has been joined by approximately 30 countries. It formulates the basic principles underlying the nuclear energy industry. Recall the basic principles: economy and competitiveness, safety and security, environmental considerations, radwaste treatment, nonproliferation, and infrastructure development. The nuclear industry must take these six principles into account. If the industry is to develop further, all these problems should be considered.

Rauf: The question has been raised, what is the problem? There is a threat that sensitive technology is spreading, which may lead to more "latent" nuclear states when political considerations change. To guard against this without a new have/have-not divide, this sensitive technology should be controlled under international auspices. We need a mechanism of assured supply, moving current facilities from national to international, and setting new facilities as international.

We have three levels: existing markets; assurance from existing market, or "virtual assurances" (including the U.S. proposal to downblend 17 tons of HEU); and a fuel bank, under IAEA auspices, that would not disturb the existing market (enough enriched uranium for one load of a 1 GW reactor)—note that the Nuclear Threat Initiative offer falls under this category. For some countries, enrichment R&D is good, leads to progress, and is not a proliferation threat. Last June the Director General said that no country should renounce its right to enrichment technology. Now there exists momentum to achieve something. The expansion of nuclear energy will call for new suppliers, new entrants, but under multilateral approaches (technology in a "black box"). Any proposal calling for a country to *forego* technology is dead on arrival. Perhaps we need a multi-vault bank—each with its own conditions. The IAEA vault would have a minimum number of conditions.

WORKSHOP – DAY 2

Tuesday, April 24, 2007

Discussion

Comments by Acting RAS Committee Chair, Academician Boris Myasoedov

This workshop has been initiated by the U.S. National Academies and the Russian Academy of Sciences. The RAS plays a special role with a lot of experience working with complex issues. The Academies of the Commonwealth of Independent States continue to interact on many issues. The president of the Ukrainian Academy believes these issues to be important, as does the Armenian Academy and the Kazakh Academy.

Comments by NAS Committee Chair, Dr. John Ahearne

There is an expansion of nuclear energy today. Countries with no NPP are considering building them, while countries with few NPPs are interested in adding more. Where will the fuel for these NPPs come from? How will it be disposed of? Will fuel be purchased, or produced? Production would require enrichment and fabrication, thus producing a proliferation threat. Uranium is available from many countries, but enrichment and fabrication is only available from a few. In addition, not too many have fabrication capabilities either. The market seems to be working.

Our South Korean participant expressed confidence that the market will continue working in the future. However, our Brazilian participant pointed out that fuel can follow a tortuous path. There is little international experience with reprocessing, nor is there much interest in the question regarding fast reactors for treatment. There seems to be trust in the IAEA, which is important for the safeguards system, especially that for enrichment. We've heard of a dormant U.S.-Egyptian agreement, which might be a model. Any new system should not exacerbate the have/have-not divide. Article IV of the NPT gives the right to various fuel cycle elements. We have seen that people are most concerned about electricity, and that nonproliferation concerns (e.g., voiced by the Director General, the U.S., and Vladimir V. Putin) may not be as widespread as believed.

If there is a nuclear renaissance, there will be a need for education and a growing number of experts. Nuclear power was a discipline in decline, and a renaissance will bring a need for new knowledgeable people. In terms of uranium, new mines are opening, which is a change from recent years when there was no incentive to explore, and some mines closed. Now uranium is at a high value. There is a need for uranium growth. Our Australian participant voiced concern that uranium-rich countries should have a central voice.

An international point: many (if not all) of the proposals have conditions. If there is a political disruption to a contract, and a country is in good standing with the IAEA, then fuel will be assured. But, should this assurance go beyond political disruptions (to include other unforeseen events)? We have seen that fuel is critical (for example, in South Korea). This raised the question of how much fuel should be stored in inventory. If there is a fuel bank, what will it be a bank of? Yellowcake? Enriched uranium (to what level of U-235 enrichment)? Fuel elements? What about spent fuel disposal? Once fuel is given, who owns it? Will it be linked to a take-back option? Permanent disposal is a problem everywhere. For new nuclear states, what will you do with spent nuclear fuel? We mustn't forget that reprocessing is also a proliferation risk. And finally, how many different options ought to be provided?

Comments by Acting RAS Committee Chair, Academician Boris Myasoedov

I agree with all of this, with one exception. The nuclear renaissance will begin soon, so this problem is very timely. Russia has a lot of hydrocarbons, has lived through the Chernobyl disaster, and has experienced a strong "green" anti-nuclear movement. We are now moving to the second stage of nuclear power revival. In 2010, we aim to commission one NPP with two units, and we wish to raise nuclear energy to 25% of our electricity by 2015. This is an historic time, in which mankind will turn to NPPs. Other sources have been discussed, but none are ready yet.

So the international community faces a problem. How can we resolve this? We must ensure, without political restrictions, access to nuclear power to all countries. How do we safeguard (in developing countries)? It would be rational to allocate funds not to their own development, but to use experience from other countries. It is important to prevent proliferation of fissionable materials. The NPT plays an important role, but this is just a treaty. Democratic countries abide by this, but can withdraw. In the Soviet Union, our political system was not a democracy. This had many negative elements, but many good things too, including the development of peaceful nuclear power, and an entire system of developing nuclear power in other countries. We had plans to supply nuclear power to other countries; they visited Soviet NPPs, and the U.S.S.R. supplied and took back fuel. We had no problems in this respect. However, this cannot be applied to today's case. These were bilateral agreements, which could be transformed in light of new circumstances.

Today, an understanding should be based on close international cooperation, without conditions, granting access to the development of nuclear energy without spent nuclear fuel treatment concerns. We should remember the spirit of the NPT, which was developed and approved despite differences of opinion

If mankind switches to a broad use of nuclear energy, we should consider a renewable approach based on fast breeder reactors and the use of spent fuel.

The Putin Initiative is a decision to create an international center. The decision has been adopted, and implementation is beginning. The following are the main ideas of the center:

1. Creation of an international center in Russia as a pilot project to provide an assured and guaranteed supply of uranium.
2. Established by countries via intergovernmental agreements to support the inalienable right to develop nuclear power without discrimination.
3. Strengthen the international nonproliferation regime using a market approach – joint enterprise (or intergovernmental agreement of other form) including nongovernmental and invited participants with no access to enrichment technologies.
4. Governments of participating states will be the executive bodies.
5. Legislative, financial, and industrial aspects will be based on existing Russian enterprise (secure provision of enrichment services, create stock of LEU).
6. Provides for regulation or access of foreign personnel to monitor quality.
7. Price of products set by co-founders, corresponding to world prices.
8. To guarantee its goals, international status is central (IAEA as an observer, under IAEA guarantees).

In conclusion, it is the only way to arrange for an international basis of nuclear energy enlargement.

Burns: From a military perspective, this discussion has been interesting and persuasive, and touched on the fundamental problems. The problem with national users of an international commodity is one of assurance. Those with their own capabilities are blessed but few. Others are dependent on shipments, and therefore are at risk of disruption. We need a broad spectrum of willing suppliers, and for users to comply with international standards.

I have a concern about the use of sanctions: I believe these are counterproductive after a short period of time. Over the long term, they no longer deprive the target; 20-25 years of sanctions does not work. A regime couched on assurance must give up long-term sanctions. Sanctions meant to punish (e.g., deprive of nuclear energy) cannot be tolerated. Multiple options to nuclear power would not be a bad thing, and a broad spectrum of options would lead to a persuasive argument.

Solonin: The further development of nuclear energy is inevitable. Furthermore, the development of fast burner and breeder reactors are necessary preconditions. The development in the near term will face problems of reprocessing. Questions of nonproliferation and the creation of dual use fissile material will be problematic in the medium term. Practically, there are 3 possible groupings: countries with all nuclear fuel cycle elements willing to develop and help with development; countries thinking of developing nuclear fuel cycle elements; countries looking to develop NPPs. All possibilities are acceptable and should be developed. The possibility of international centers is not yet there. We are still working on principles, but have nothing yet in practice. We should take practical steps to convince others to use this possibility, then see if it works or not. This discussion is very useful.

Gottemoeller: The previous speaker's comments were very useful. Developing a more international nuclear fuel cycle requires broad acceptance. We must convince others over time to cooperate. This will require proof of practical utility. We must take measured steps with a pilot program in order to build confidence.

Regarding the issue of incentives as opposed to sanctions: How do we provide more incentives? New technologies? The possibility of technical cooperation (new fuel, reactors)? This is now restricted to capital-rich countries, others are left on the margins. In the future, it would be worth considering drawing countries into a new approach on the front or back end?

Bezzubtsev: As a representative of radiation security and regulation, I am interested in international requirements for safety and the normative base of development of more nuclear energy. The question of safety is important, and a national base is required to harmonize safety internationally; individual countries will develop their own rules and regulations.

Yesterday, most of the discussion was on nuclear fuel centers and enrichment services. Countries interested in nuclear energy are not interested in UO2 or UF_6. Rather, they are interested in fuel assemblies. Therefore, they may opt for leasing services through international centers, in order to get the final product (meeting security requirements) and to train staff to use irradiated fuel. Enrichment services and the market will require a normative base, which Russia has. Russia has just passed a law on fissile material ownership. More attractive would be a center intended to handle irradiated fuel. In Russia, nuclear fuel can be repatriated for long-term storage or reprocessing–this is attractive to new countries. There is a legal base for handling of irradiated fuel, and Bulgarian and Ukrainian fuel is sent back for reprocessing or storage. Over the past two years, there has been a joint U.S./Russian repatriation project (for research fuel) under Russian rules for safe handling and addressing ecological problems. The radwaste problem is a big concern. It will be important to take back and store some waste. A law is being prepared to resolve pending issues of radwaste handling as Russia cannot now accept radiological waste from other countries.

Levenson: From the U.S. perspective, I cannot speak of ongoing projects. Few if any countries rank proliferation concerns over energy security. Neither the haves nor the have-nots will forego anything in the long term (e.g., the United States first decided against reprocessing then changed its mind). There are two groups (open fuel cycle and closed) with two different time schedules and should be considered separately. In the near term, the focus will be on an open cycle: assured quantity of fuel, timely delivery, and at a good price (the "carrot" system). Countries will occasionally reassess domestic development, therefore an "assured supply" must remain economical, i.e. must remain cheaper than a domestic supply. Energy security is so important that even with an assured supply countries may want alternatives.

There are three parts of an assurance of supply: assured quantities, schedule, and price. Price might be the most important. Countries with expanding nuclear programs will assess whether a domestic program is cheaper. The central system should not be so profit driven that they ignore this. Even if a country is a participant in a multinational supply, they will probably also seek other arrangements for security of supply system.

In terms of a closed cycle with advanced technologies, in the past, there has been lots of nuclear power though all was based on Canadian, Russian, or U.S. technology. Domestic technology was not necessary. We should make it clear to users that technology will be shared with "have-nots" wishing to move to a closed fuel cycle. When a country is ready to move to a closed fuel cycle, they should be assisted in doing so.

Energy security, nonproliferation, and economics sometimes compete; we need a middle ground.

Bychkov: I would like to touch on several parallel issues: supply of fresh fuel and reprocessing; transportation; waste storage; long-term isolation and immobilization; and accidents. There must be an overall system of services not just supplies. Then, an approach to internationalization would not be a copy of current systems, there would be a need for further development.

Perhaps we can optimize the development of the nuclear fuel cycle. For example, consider a country with several fast reactors that need the fissile material for operation; that country would undertake reprocessing. Now, the fuel cycle is focused on enrichment, this will shift to joint R&D and joint systems (e.g., fast reactors). In theory, these are components of GNEP, but there is still a long way to go. From practical experience in international initiatives (e.g., INPRO), we have developed new ways to oversee the spread of technology and its transfer. This knowledge would tend to make possible the oversight of certain nonproliferation facilities. A super-national mechanism would work as an incentive to improve technology.

Fetter: Growth in nuclear power is not inevitable, but desirable (because it can have no CO_2 emissions), but it must grow by a factor of five to make a dent in overall CO_2 emissions. If this leads to a number of countries able to enrich and reprocess, there will be many virtual nuclear weapon states and this is not desirable. This latent capability will create tension and distrust, which may outweigh benefits. If there is an expansion, it will be necessary to limit the number of countries with enrichment and reprocessing technologies. The idea of international centers is interesting, but limited. A fuel bank can play a small role.

My impression from the presentations is that waste is one of the biggest concerns. Therefore, if there were a central, international option for long-term storage, take back, and disposal, this would be a good incentive. A leasing and take back mechanism may be the most promising. More countries may wish that we consider take back (though not necessarily with reprocessing, at least not until economically necessary). It would be good if more countries were open to hosting an international spent fuel storage or disposal facility.

Petrov: There will be a need to train future experts to implement what we are now discussing. We have paid close attention to uranium mining and technologies, but not to the waste (which will be a result of these activities). Access to the knowledge market is not free to developing countries. It will be up to states to resolve these issues.

There will soon be a problem with mining training. With respect to training in waste disposal, if problems, obstacles, and difficulties are not resolved, the accumulated spent nuclear fuel will serve as a brake on future development, as is an international center for training in "subsoil sciences" to set up underground repositories. This should be a principle issue. There is a need for a center with international experts of geology, and other fields for a future international spent fuel center (under the IAEA).

Budnitz: I am not convinced that the number of countries, whose participation in an assured fuel scheme would be crucial to improved nonproliferation, is sufficient. Consider this "Berkeley metaphor." The city of Oakland has one of the highest murder rates in the country, and therefore so does Alameda County. However, if I convinced all my friends to sign a pact saying that they will not commit murder, this would have very little effect on the murder rate in Alameda County. Would such a scheme change the behavior of countries that would otherwise misbehave? However, the scheme is worth pursuing, because I may be wrong. My question is this: Does a large number of countries, whose participation would make a difference, exist? Once a country is determined to misbehave, this scheme will be of little help. No country will ever give up an inalienable right.

Bunn: If we want to grow nuclear energy (over the next 40-50 years) without increasing proliferation risks, we must limit enrichment and reprocessing technology. We need to provide an incentive to countries to make a sovereign choice to not develop enrichment and reprocessing. The commercial market seems to work well. In this discussion, several classes of countries have been mentioned:

1. Some countries are not interested in enrichment or reprocessing technologies, so this scheme won't matter to them.
2. Some countries are interested in enrichment, but assurances of supply are not important (e.g., Australia, Brazil).
3. For some countries, enrichment was used to pursue nuclear weapons.
4. There is a set of countries "on the fence." If more assurance is given than the commercial market, this may affect internal debates.

For this last class, the idea is worth pursuing. However, we must not "mess up" the commercial market. There is a distinction between a fuel bank (no sign up) and an international fuel cycle center (sign up and pay). Must a country make a decision right now? Can they do so in 15 years? What are the conditions for assurance or denial of fuel? Forego forever? (Not realistic.) Refrain from enrichment while participating? Good nonproliferation standing? (Who decides? UN Security Council? IAEA? The U.S. cutoff of Brazil is perceived differently in Brazil and the U.S.) If suppliers could overcome the political obstacles to take-back, and we could find a host for an international facility, this would be a major breakthrough for nuclear energy. Fresh fuel supply (front end) works well, but spent nuclear fuel handling (back end) is more problematic. If we could offer a country both services, we would have lots of leverage. We can applaud Russia for doing this with Iran and potentially other countries.

Galstyan: During this meeting we have heard a wide range of views, and I respect the different positions. Armenia has a pragmatic position and prospects are assessed realistically. For example, there is no infringement on our pride that commercial airliners are produced by only two companies, Airbus and Boeing. We understand that some technologies are only realistic for certain rich countries.

Our big concern is the following: What about spent nuclear fuel? Armenia understands what it means to lose electricity security and to lose NPPs. Despite good fuel contracts, there are no 100% guarantees about hydropower or fossil fuels. Nuclear energy is a "delicate sphere" calling for different fuel schemes (different from those in the common market) taking into account dual use technology.

Villagra-Delgado: Article VI of the NPT is important, and should not be put aside. Regarding (bilateral) assurances of supply and multilateral nuclear approaches, we must ask, what is the objective? Enhanced nonproliferation? These approaches will add little as they will apply only to well-behaved countries as the others will not join them. A revival of nuclear energy? This might be useful.

We must recognize that the safeguards system is working well. We have had no new problem countries in the NPT regime for more than 20 years. On the other hand, countries outside of the NPT framework, which have developed nuclear weapons are in the process of getting a "clean bill of health." Is that not a contradiction? I conclude that effective safeguards are the way to go.

How are assurances presented? What are the conditions for guarantee of supply? If the safeguards system is not working, if the Additional Protocol is so flawed, why should a country sign onto it? Stating that they are not good enough and that we need something totally new may be self-defeating and weaken the whole regime. Pursuing an assurance of supply may send the wrong message to countries about the functioning of the market. The "haves" telling the "have nots" is too complicated and too expensive; this is unacceptable. Saying that the enrichers are virtually proliferators presumes that the countries want to proliferate instead of just developing civil nuclear energy, which is the reality in most cases.

Then there is the question of "breakout." I disagree entirely that this is unilateral legal option. If a country uses its membership in the NPT to develop nuclear weapons, it would have acted contrary to its obligations under the Treaty and would be in breach of it. Upon declaring its intention to opt out of the Treaty, the matter must be referred to the UN Security Council where action could be taken. Agreement in the Council, particularly among the P5 countries will be needed. If these cannot agree on such a case, there is no multilateral approach that could solve that lack of will to act. Besides, it is clear that an assurance of supply based on a "join and renounce or else" system will terminate the commercial market and could be perceived as an attempt at the cartelization of it.

dos Santos Guimarães: I pose the question: What is the problem? It seems there are two routes to nuclear fuel: purchase turnkey fuel assemblies or purchase uranium and fuel services. Fuel assemblies are customized commodities. Even with our cooperative development with Korea and Slovenia, we cannot just ask them to provide us with some assemblies. For turnkey NPPs, an international fuel bank will not provide assurances. For uranium and fuel services, this may be useful (e.g., the Russian proposal) but we still have the problem of fuel fabrication. A country must have its own fuel fabrication facility. Therefore, a take-back option would be very interesting. This should be universalized.

Power should be transferred from utilities to fuel providers. This would have a non-marginal impact on nuclear power economics. In regard to reprocessing, Brazil has no interest. In general, the problem lies with research reactors. Reactor grade plutonium is infeasible for use in a weapon, therefore reprocessing of light water reactor fuel is not a problem. The only concern is that if you can reprocess one kind of fuel then you can reprocess any kind. In regard to enrichment technologies, this technology was born in Germany and then escaped (both diffusion and centrifuge). Our experience is that we made some batches of 20% U-235 for research reactors under full safeguards (with lots of inspections). Above 10-12%, we have problems with hexafluoride impurities. Additional purification of hexafluoride is needed to go up to high enrichment, and that additional purification is easily detected by safeguards. Impurities could be introduced for these reasons. We can discuss these issues directly.

We should discuss "haves" and "have notes" in the open, why do this implicitly with subterfuge calling it assurance of supply? Throughout history, energy technology has changed mankind (as the level of "energy concentration" increased): first wood for fire, then coal, then oil, then gas, then uranium. The question is who will control uranium? The United Kingdom became predominant as wood turned to coal, and the United States became predominant as coal turned to oil. As oil turns to uranium, who will take control? Consider the question of assuring supply. The 18th century United Kingdom Portugal Agreement assured a supply of manufactured products in exchange for olives and wine, and as a result Portugal is underdeveloped to this day.

Ahearne: I have to point out that a past NAS study on plutonium disposition concluded that reactor-grade Pu can be used for a crude but powerful weapon.

Smith: Proliferation is a political, not a technical problem. The common market works well for supply. We must separate "multilateral" from "multinational." The multinational nature of the commercial enrichment market is an important element of assurance of supply. If Australia is to enrich, we should join Urenco. Any mechanism to guarantee enrichment should not interfere with the common market. Reprocessing is a different question: it is more dangerous for proliferation and there is not a broad market. Without this, nuclear power is wasteful (only a small percentage of energy is extracted). There is an opportunity for a multilateral closing of the fuel cycle, in a manner to maximize energy extraction while minimizing waste. I believe we are looking at the wrong end: Enrichment is working, and the true opportunities lie with reprocessing.

Shaker: I call attention to the Expert Group's report. If we want to internationalize the nuclear fuel cycle, this should be done gradually and built on confidence, step-by-step. It cannot be done universally (neither in participation nor in the parts of the fuel cycle being internationalized). Currently, there are no problems in NPP countries, but newcomers may need help with supply. (A history of colonialization and deprivation leads to a desire for assurance.) But, do we guarantee fuel? Enriched uranium? Natural uranium? For example, to an automobile owner, which is more important, guaranteed access a gas station or an oil well? It will be important in the future to have multiple supply centers, ensuring competition and an improvement in quality, delivery, and price.

Regarding the NPT: If a country withdraws, why? The UN Security Council is the judge. We should leave the withdrawal clause in place–it is the safety valve. It is important to give newcomers a chance to have their say. They should participate in decision making. We should not divide further into "haves" and "have-nots." We should give future importers a role. Urenco and Eurodif should allow the participation of others. Brazil and Japan have the potential to become regional centers.

Stamenov: On what basis should we draw our conclusions? Do we want to increase energy capacity and diversity while reducing emissions? Do we want to improve conservation? Our

2004 decision for 2 new VVER-1000 units affirms our choice for the next 30-40 years. Taking into account that all experiments to put other kinds of fuel in the VVER were not exactly successful, we are committed to Russian production.

The problem seems to lie with spent fuel—transportation, repatriation, and storage. If there would be an international center for spent fuel, Bulgaria would be very interested. If you have limited expert resources (a small country) then you can rely on a larger country to address those nuclear problems and focus on other problems. We have excellent experience in fresh fuel and spent nuclear fuel transportation, and have solved some ecological and social problems. Transportation is a serious problem. Bulgaria is interested in an international spent nuclear fuel solution. From a legal point of view, there is a difference in view: is spent nuclear fuel "waste" or "material?" Experience with international centers suggests that if they start on a commercial basis, there are many potential problems. It is better to begin work from a budget basis from the participating countries.

Yang: How would an international center be managed? Would it be multinational, or be managed by the IAEA? What kinds of contracts? Commercial or intergovernmental? What about export regulations? There is not much difference between government contracts and commercial arrangements because everyone needs an export license from the government.

Karyono: In Indonesia, we have no experience with NPP operation, only with fuel fabrication and research reactors. We believe that the R&D industry has a duty. What is the optimum solution? Regarding a safeguards system, would this be multinational, IAEA, or based on bilateral MOUs?

McDonald: First, I believe there is a semantic confusion, between "assurance of supply" and "assurance against interruption." There is a general question of assurance—international mechanism to facilitate the nuclear renaissance—against political or commercial interruptions. The IAEA will restrict our attention to political interruptions only. No mechanism should disrupt the commercial market. Second, I'd like to call for an "upbeat intervention." We must be careful to respect everybody's rights. Nobody wants to give up any rights. But consider the Dutch/German/U.K. proposal for "enrichment bonds:" a guarantee to produce, enrich, and provide export licensing for uranium. This is a big step. They are ceding some of their sovereign rights to the IAEA for the greater good. This proposal was voluntarily put forth, and is a big step. Our opinion from the Angarsk visit, is that this is a very pragmatic approach.

Third, I pose the following "stumper." Consider a country like Australia. It is not worried about political risk, and considering commercial activities. The IAEA assurance mechanism is not a priority, but they certainly have no objection. In the end, it may even help them. Article IV of the NPT requires us to act in a non-discriminatory manner. But, Australian export controls are deliberately discriminatory. If Australia signs on, then the IAEA plan becomes discriminatory. Would it be better not to include Australia?

Starz: (A) Why now? These ideas have been around for a long time. Yet, some things are more obvious now (highlighted by the nuclear renaissance). There is a new concern for the proliferation of sensitive technologies (and so called "latent" capabilities). Now, momentum is building (from suppliers and proposals). Will it lead to something this time? (B) What is the rush? The goal is to reduce the incentives for domestic nuclear fuel cycle capabilities. Do we have time to do so? If not, why not? (C) It seems that the beneficiary pays. What will be the conditions? Will they stick? Those with the highest benefits should have the least cost in giving up their rights. This may help sort out different options.

Kelly: I am interested in the idea of incentives, and a focus on the back end. This could free countries from their own regulations. We should have a system to incorporate storage of spent nuclear fuel into these front-end discussions. We also need training in final spent nuclear fuel management. How could this benefit from an international partnership?

Shalabi: What seems to be missing are political options for take-back and leasing. Going back to Ambassador Shaker's automobile analogy, why not put a fuel cell right in the car? Finally, what is meant by "discriminatory?" This is a regulatory decision and not something else.

McDonald: Australians have explained that a discriminatory process (the right is ultimately reserved to say no) is distinct from a criteria-based approach (if all criteria are met, ok). In Russia, for example, if Russia is a supplier and the IAEA agrees to supply, does Russia have to agree? Shouldn't we just combine Russian and IAEA criteria from the beginning?

Berriman: In Australia, our policy is not "discriminatory," but "selective."

Closing Remarks by Acting RAS Committee Chair, Academician Boris Myasoedov:

During this workshop we have worked interactively, which has been very important. Of course we didn't solve all problems, and more seminars are needed. We have heard some controversial opinions, which should be studied. If international centers are created, should there be many? Few? What about geographical considerations? We should work together to avoid conflicts. These centers should deal not only with enrichment approaches, and it should be a gradual process.

We should not use a natural uranium bank, and we should consider spent nuclear fuel storage problems. All steps should be voluntary. We shouldn't push the process and put all countries together into "one small room."

Russia has made some practical steps. The Russian system was developed independently, but there is a need for cooperation. An international system is not needed for Russia alone, we have good experience in all steps of the nuclear fuel cycle, including fast reactors. Instead, we need to combine our efforts.

LIST OF WORKSHOP PARTICIPANTS

Argentina
Villagra Delgado, Pedro Raul, Embassy of Argentina to Australia
Wilkinson, Moira, Permanent Mission of Argentina to the IAEA

Armenia
Galstyan, Areg, Ministry of Energy of Armenia

Australia
Berriman, Annette, Department of Foreign Affairs and Trade of Australia
Hutchings, Ronald, Permanent Mission of Austria to the IAEA
Smith, Ian, Australian Nuclear Science and Technology Organization
Walker, Martin, Permanent Mission of Australia to the IAEA

Brazil
Guimarães, Leonam dos Santos, Eletronuclear
Vallim Guerreiro, Antonio José, Permanent Mission of Brazil to the IAEA

Bulgaria
Stamenov, Jordan, BAS, Institute for Nuclear Research and Nuclear Energy

Canada
Shalabi, Ahmed, Permanent Mission of Canada to the IAEA

Egypt
Shaker, Mohamed, Egyptian Council for Foreign Affairs

Indonesia
Hiswara, Eri, Permanent Mission of Indonesia to the IAEA
Karyono, H.S., National Nuclear Energy Agency (BATAN)

Korea, Republic of
Cho, Chang-Sok, Korea Nuclear Fuel Company, Ltd.
Yang, Chang-Kook, Korea Nuclear Fuel Company, Ltd.

Russian Federation
Bezzubtsev, Valery S., Federal Service for Ecological, Technological and Atomic Supervision of the Russian Federation
Bychkov, Alexander V., Federal State Unitary Enterprise "State Scientific Center of Russian Federation"; Research Institute of Atomic Reactors
Ivanov, Valentin B., Member of the State Duma Russian Parliament
Lushnikova, Liudmila, Institute of Biochemical Physics, Russian Academy of Sciences
Myasoedov, Boris F., Russian Academy of Sciences
Petrov, Vladislav A., Institute for Ore Deposits Geology, Petrography, Minerology and Geochemistry of the Russian Academy of Sciences
Shiyan, Yuri K., Russian Academy of Sciences
Solonin, Michael I., TVEL Corporation
Zmeyevsky, A., Permanent Mission of the Russian Federation to the International Organizations in Vienna

Switzerland
Kelly, Julian, Association for Regional and International Underground Storage (ARIUS)

United States of America
Ahearne, John, Scientific Research Society (SIGMA XI)
Budnitz, Robert J., Lawrence Livermore National Laboratory
Bunn, Matthew, Harvard University
Burns, William F., Major General (USA, retired)
Fetter, Steve, University of Maryland
Gottemoeller, Rose, Carnegie Endowment for International Peace
Guenther, Rita, The National Academies
Humphrey, Marc, The National Academies
Levenson, Milton, Bechtel International (retired)
Lowenthal, Micah, The National Academies
Starz, Anne, Permanent Mission of the USA to the UN Organizations in Vienna

International Atomic Energy Agency
Forrstroem, Hans, Department of Nuclear Energy
McDonald, Alan, Department of Nuclear Energy Planning and Economic Studies
Rauf, Tariq, EXPO —Verification & Security Policy Coordination
Simpson, Fiona, EXPO—Verification & Security Policy Coordination

APPENDIX C

THE STRATEGY OF NUCLEAR ENERGY DEVELOPMENT IN RUSSIA

By Alexander Bychkov

Nuclear power engineering development is envisioned as an integral part of the Russian Federation's energy strategy, and Russia is now taking several steps to further develop and expand its use of nuclear power. Russia is investing in construction of several nuclear power plants, including both pressurized water reactors (VVERs) and a liquid-metal cooled fast reactor (the BN-800). A prototype of a compact, portable (floating) nuclear power plant has also been built. Russia has also "reconstructed and reorganized its whole nuclear enterprise, consolidating and reorganizing nearly all of the nuclear functions into a state-owned corporation. In addition, the nation is working to develop new reactors and new closed fuel cycles.

The strategy for nuclear power engineering development in the first half of the 21st century is based on the following principles: nuclear fuel breeding, comprehensive safety, and competitiveness. Projections vary, but energy demands in Russia are expected to increase by 50 percent between 2006 and 2016, and rise to double the 2006 level by 2020. Demand for electricity is expected to grow more slowly (rising 50% by 2020, and 100% by 2030, compared to 2005 levels), but still steadily and strongly. *The Strategy for Development in the Russian Nuclear Power Sector from 2007 to 2015* provides for implementation and growth of several federal programs, as well as enacting of the law on restructuring of the civil branch of the nuclear power sector, which was passed in early 2008. The Rosatom Corporation is now established and establishment of Atomenergoprom is to be completed in 2008, thus incorporating all parts of nuclear manufacturing cycle, from uranium mining and enrichment, reactor design and construction, and power plant design, construction and operation.

Nuclear power's contribution to the energy strategy can be achieved through several investments for near-term and longer-term results. In October 2006, the Russian Federation accepted the Federal Task Program "Development of Russia's Atomic Power Complex from 2007 – 2010, and to 2015." This spelled out the directions of nuclear power development into the future: (1) development of nuclear power capacities, (2) development and renovation of fuel cycle capacities, (3) development of capacities on management of spent nuclear fuel and radioactive wastes of nuclear power plants and preparation of nuclear reactors for decommissioning, and (4) transition to innovative nuclear technologies.

For very near-term results, the current set of nuclear power plants can be maintained and operated more effectively, including upgrading and extending the lifetime of the operating power units; increasing their efficiency and maximum utilization of capacities (load or capacity factor);

and design and construction of spent nuclear fuel and radioactive waste facilities so that power-plant operations are not inhibited by accumulation of these materials.

At present, 31 reactors operate at 10 nuclear power plants in Russia (see Figure C-1 for the locations of these and future nuclear power plants). Beginning in 2007, each year Russia plans to initiate construction of at least two nuclear power units with a combined capacity of about two gigawatts electric (GWe). By 2015, the Russian Federation plans to invest approximately 1.5 trillion rubles in the design and construction of new NPPs. If this schedule is kept, 10 new nuclear power reactors with an installed capacity of 9.8 GWe will be put into operation by 2015, raising the total nuclear generating capacity in Russia from its current level of 23.2 GWe to 33 GWe. This would increase the nuclear power share of Russia's nuclear generating capacity to an estimated 18.6%. Beyond 2015, the plans are even more ambitious: construction of between three and four nuclear power units annually. By 2030, the goal is for nuclear power plants to generate 25% of Russia's electricity. Figures C-2 and C-3 illustrate the planned growth. With such significant expansion of its use of nuclear power, Russia has concluded that it should develop a systematic solution to problems concerning spent nuclear fuel and radioactive waste.

FIGURE C-1 Locations of existing and planned future nuclear power plants in the Russian Federation.

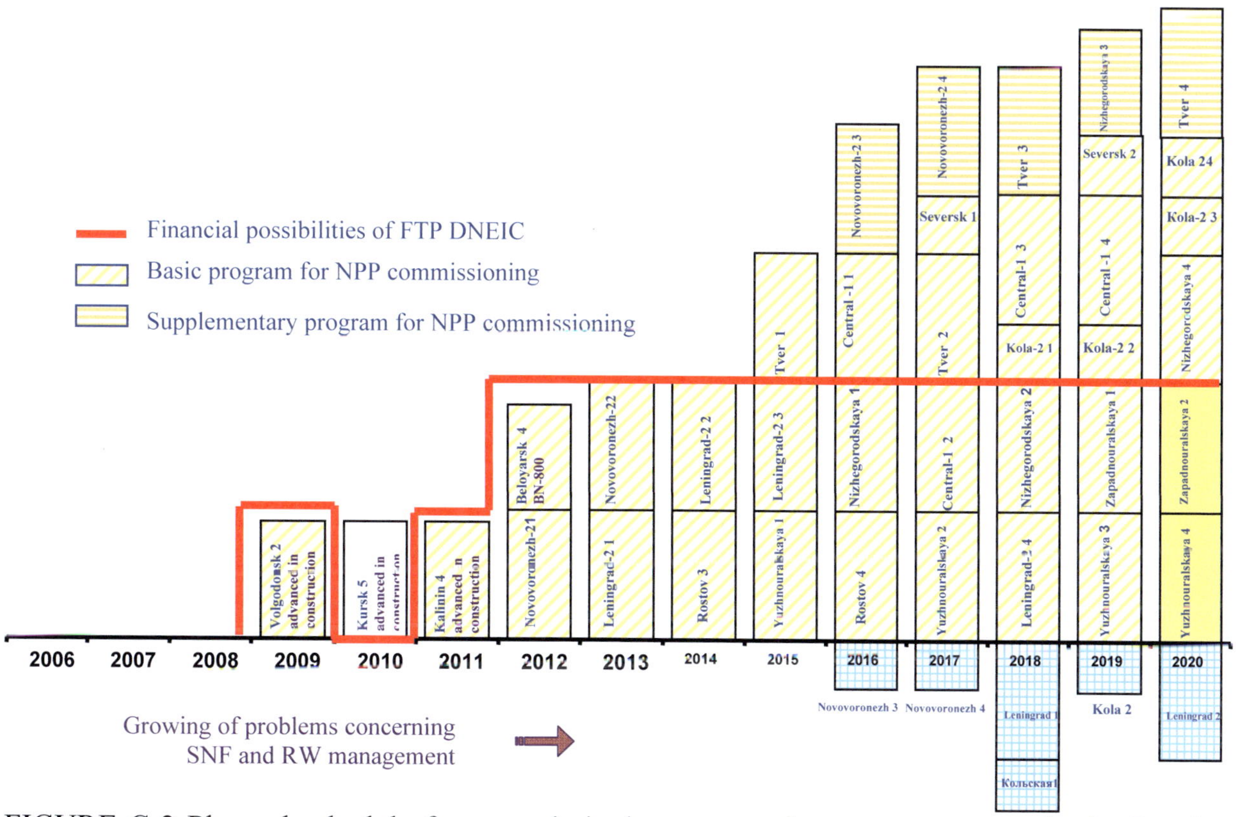

FIGURE C-2 Planned schedule for commissioning new nuclear power reactors in the Russian Federation.

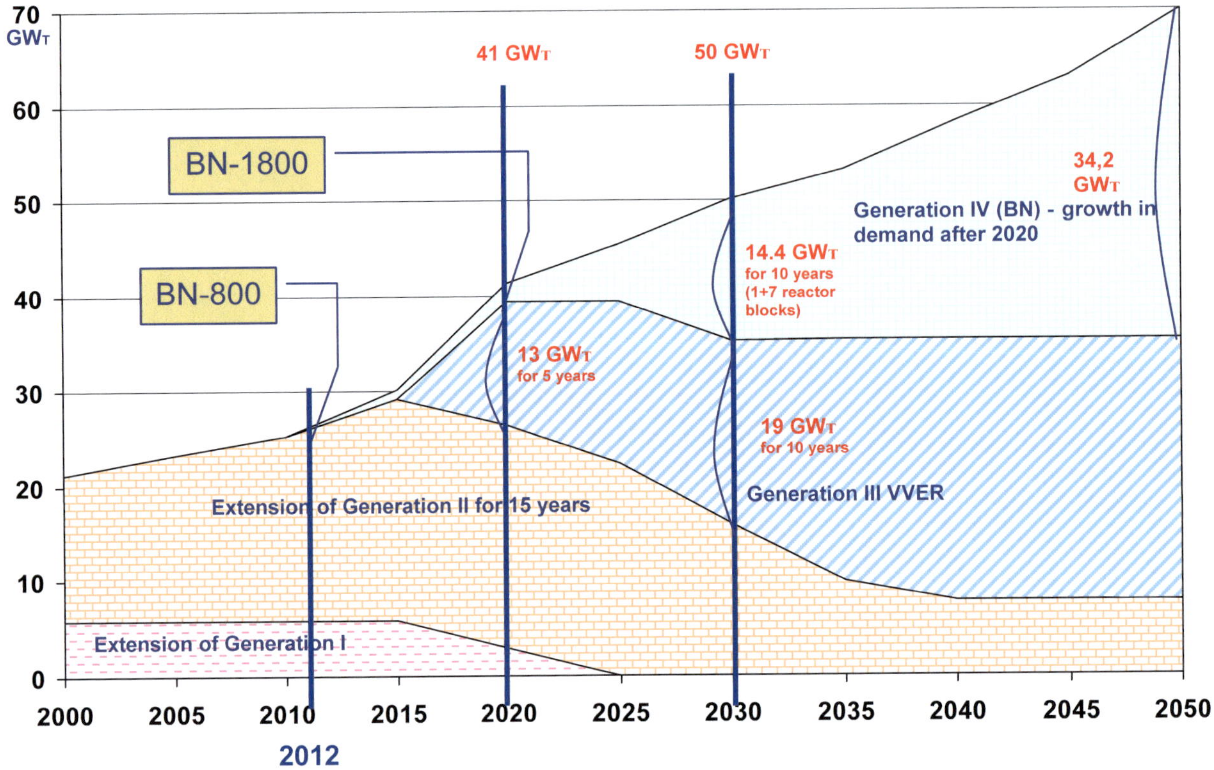

FIGURE C-3. Planned contributions of different reactor types to nuclear power generation in the Russian Federation through 2050.

Applying fast reactor technology with a closed fuel cycle extends the resource potential of the nuclear power fuel supply. Future nuclear power engineering can develop based on fast-reactor technology. Russia has unique experience in the development and operation of fast reactor nuclear power plants: 20-year successful operation of the BN-350 and operating BN-600 unit 3 at Beloyarsk Nuclear Power Plant.

Russia is now building on its experience with fast reactors by starting a closed fuel cycle with the BN-600 and BN-800 reactors. Preparation of a hybrid core for the BN-600 with MOX fuel was initiated in 2007. This plan is illustrated in Figure C-4. Production of MOX fuel for the BN-800 is planned to begin in 2011, a year before the BN-800 is scheduled to start up. In the period from 2016 to 2018, Russia's plans call for implementation of semi-industrial BN-800 closed fuel cycle technologies (see Figure C-5). A fully industrial-scale fast reactor with a closed fuel cycle is planned between 2018 and 2020.

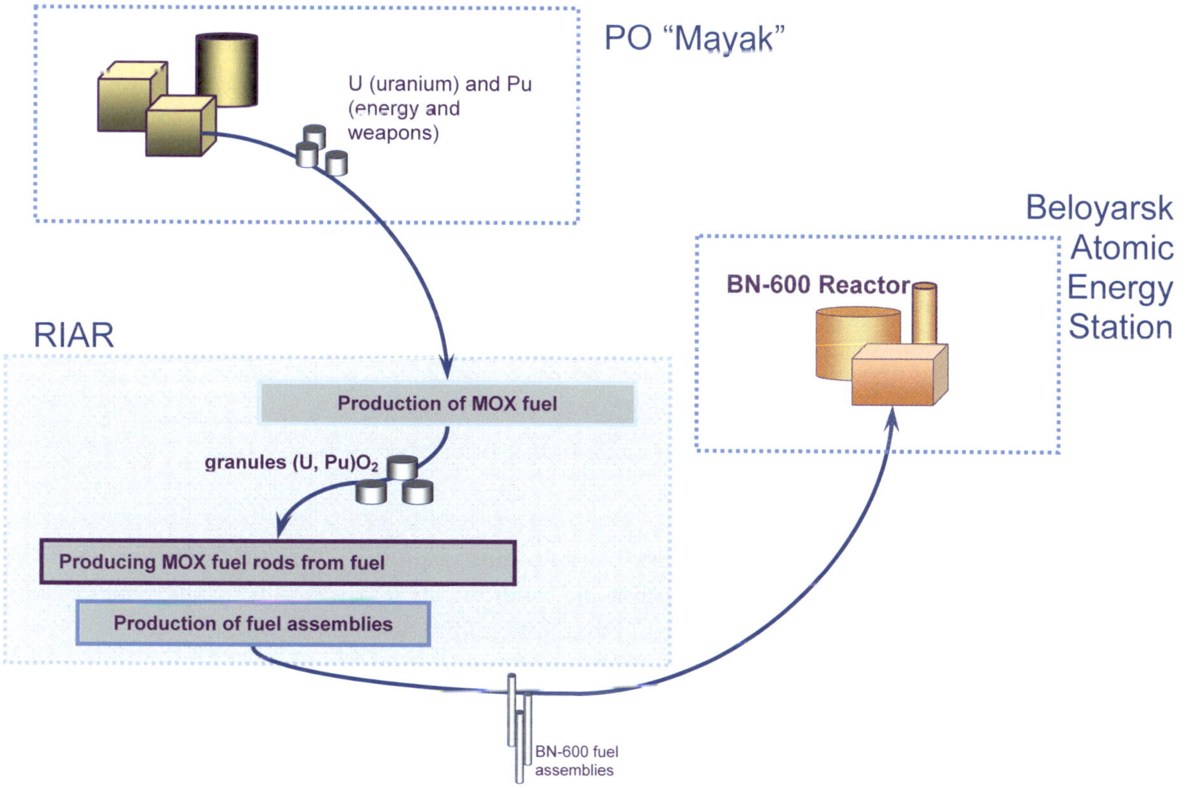

FIGURE C-4 Planned fueling of the BN-600 reactor using highly enriched uranium and separated plutonium already in storage.

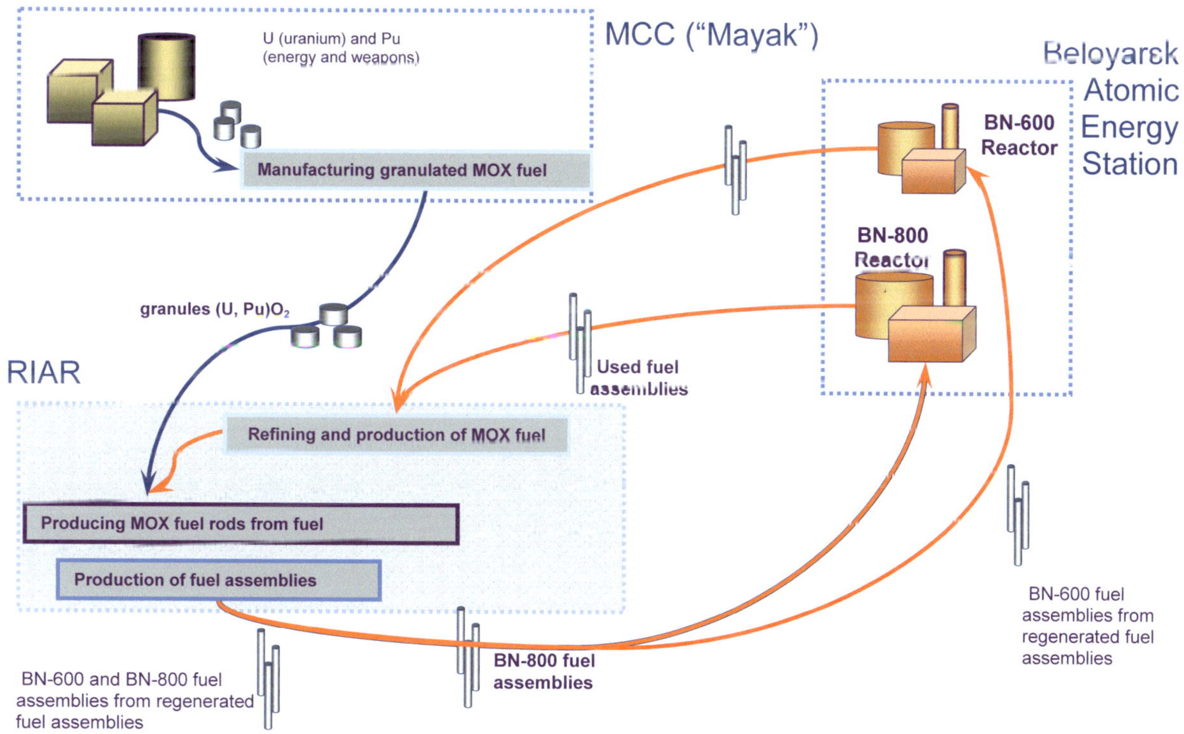

FIGURE C-5 Planned closed fuel cycles involving the BN-600 and BN-800 reactors.

In late 2007, several key decisions were made regarding Russia's future fuel cycles. First, it was decided that MOX-fuel production would be based on pyroelectrochemical methods and vibropacking moving toward closing the fuel cycle with compact, dry technologies for recycling spent nuclear fuel and simplified technologies of fuel-pin manufacture, developed at the Research Institute of Atomic Reactors (RIAR). The goals for the closed fuel cycles under development are: minimization of expenses for spent fuel recycling, fuel pins refabrication and waste treatment; minimization of radioactive waste volume and complete recycle of minor-actinides for transmutation in the same system; exclusion of pure fissile materials (plutonium) from recycling technologies; and arrangement of all procedures in remote systems.

To assist with the development of new fuel cycles, new facilities and activities are planned. These include design and construction in Dimitrovgrad of a new, multi-functional fast test reactor–sodium cooled with autonomous loops–for testing of fuels, materials and technologies.

Pilot and industrial facilities for fuel production (including MOX fuel) and investigations of fuel cycle processes (test-demonstration centers for aqueous and dry processes) will be created. Generation IV demonstration reactor systems are also planned under the New Federal Task Program from 2008 ("Nuclear Energy Technologies of New Generation").

APPENDIX D

AGREEMENT BETWEEN THE GOVERNMENT OF THE UNITED STATES OF AMERICA AND THE GOVERNMENT OF THE RUSSIAN FEDERATION FOR COOPERATION IN THE FIELD OF PEACEFUL USES OF NUCLEAR ENERGY

The Government of the United States of America and the Government of the Russian Federation, hereinafter referred to as the Parties;

Convinced that the use of nuclear energy for peaceful purposes is a reliable basis for meeting national energy sector requirements in a manner that is sustainable, environmentally safe, and economically beneficial;

Seeking to expand and enhance mutually beneficial cooperation in the field of the peaceful uses of nuclear energy on a stable, reliable, and predictable basis;

Recognizing that the expansion and enhancement of cooperation between the United States of America and the Russian Federation on an equal footing will help strengthen international stability, as well as promote political and economic progress;

Taking into account that both the United States of America and the Russian Federation have achieved an advanced level in the use of nuclear energy for production of electric power and in the development of nuclear industry and scientific research in this field, and guided by the common goals of achieving a higher level of safety and protection of populations and the environment;

Mindful of their respective obligations under the Treaty on the Non-Proliferation of Nuclear Weapons of July 1, 1968 ("NPT"), to which both the United States of America and the Russian Federation are parties;

Reaffirming their commitment to the international development and use of nuclear energy for peaceful purposes that are consistent with the provisions of the NPT;

Taking into account that the United States of America and the Russian Federation are members of the International Atomic Energy Agency ("IAEA");

Affirming their support for the objectives and Statute of the IAEA and their commitment to the Guidelines of the Nuclear Suppliers Group;

Acknowledging the importance of the provision of nuclear fuel supply assurances under the auspices of the IAEA;

Acknowledging the need for measures for the physical protection of nuclear material and facilities and affirming compliance with the obligations set forth in the Convention on the Physical Protection of Nuclear Material of October 26, 1979, to which the United States of America and the Russian Federation are parties;

Expressing a firm commitment to strengthening the international regime of nuclear non-proliferation and IAEA safeguards;

Noting the need to establish conditions governing the transfer for peaceful purposes of nuclear material, relevant equipment and technologies between the United States of America and the Russian Federation that avoid interference in the civilian nuclear programs of the United States of America and the Russian Federation;

Mindful that peaceful nuclear activities must be undertaken taking into account the need to ensure protection of the international population and environment from radioactive, chemical and thermal contamination;

Have agreed as follows:

ARTICLE 1

For the purposes of this Agreement, the terms listed below shall have the following meanings:

1. "Component" means a component part of equipment or other item so designated by agreement of the competent authorities of the Parties;

2. "Equipment" means any reactor, other than one designed or used primarily for the production of plutonium or uranium-233, or any other item so designated by agreement of the competent authorities of the Parties. "Reactor" means any apparatus, other than a nuclear weapon or other nuclear explosive device, in which a self-sustaining fission chain reaction is maintained. The phrase "designed or used primarily for the production of plutonium or uranium-233" shall not apply to breeder reactors that do not produce nuclear material for use in nuclear explosive devices, nor with respect to reactors primarily used for the production of plutonium-238;

3. "High enriched uranium" means uranium enriched to twenty percent or greater in the isotope uranium-235;

4. "Information" means scientific, commercial or technical data or information in any form that are appropriately designated by agreement of the competent authorities of the Parties to be provided or exchanged under this Agreement;

5. "Low enriched uranium" means uranium containing less than twenty percent of the isotope uranium-235, but more than the content ofuranium-235 in natural uranium;

6. "Major critical component" means any part or group of parts essential to the operation of a sensitive nuclear facility;

7. "Moderator material" means heavy water, or any other material suitable for use in a reactor to slow down neutrons and increase the likelihood of further fission, as jointly designated by the competent authorities of the Parties;

8. "Nuclear material" means source material and special fissionable material, and includes, *inter alia,* irradiated source material and irradiated special fissionable material. "Source material" means uranium containing the mixture of isotopes occurring in nature; uranium depleted in the isotope uranium-235; thorium; any of the foregoing in the form of metal, alloy, chemical compound, or concentrate; any other material containing one or more of the foregoing in such concentration as the Board of Governors of the IAEA shall from time to time determine; and such other materials as the Board of Governors of the IAEA shall from time to time determine or as may be agreed by the Parties. "Special fissionable material" means plutonium, uranium-233, uranium enriched in the isotopes uranium-233 or uranium-235; any material containing one or more of the foregoing; and such other fissionable material as the Board of Governors of the IAEA shall determine or as may be agreed by the Parties. "Special fissionable material" does not include "source material." Any determination by the Board of Governors of the IAEA under Article XX of the IAEA Statute or any determination by the Board of Governors of the IAEA that otherwise amends the list of materials considered to be "source material" or "special fissionable material" shall have effect for the purposes of this Agreement only when the Parties have informed each other in writing that they accept this amendment. For the purposes of this Agreement, "plutonium" does not include plutonium with a content of the isotope plutonium-238 exceeding eighty percent;

9. "Peaceful purposes" or "peaceful use(s)" include the use of information, nuclear material, moderator material, equipment and components in such fields as scientific research, electric power generation, medicine, agriculture and industry, but do not include their use in, or use for research on or development of, any nuclear explosive devices or any military purposes. Military purposes shall not include provision of power for military bases drawn from any power network, production of radioisotopes to be used for medical purposes in military hospitals, and other similar purposes as may be agreed by the Parties;

10. "Authorized person" means any individual subject to the jurisdiction of the United States of America and any legal entity, including a joint venture or partnership, subject to the jurisdiction of either Party, that is authorized by the relevant Party to implement cooperation under this Agreement, but does not include the Parties to this Agreement;

11. "Restricted Data" means all data concerning (1) design, manufacture or utilization of nuclear weapons, (2) the production of special fissionable material, or (3) the use of special fissionable

material in the production of energy, but shall not include data that the Government of the United States of America has declassified or removed from the category of Restricted Data;

12. "Russian Federation State Secret Information" means information protected by the Russian Federation in the area of its military, foreign policy, economic and other activities, whose dissemination could be detrimental to the security of the Russian Federation;

13. "Sensitive nuclear facility" means any facility designed or used primarily for uranium enrichment, reprocessing of irradiated nuclear material, heavy water production, or fabrication of nuclear fuel containing plutonium;

14. "Sensitive nuclear technology" means any information, including information that is incorporated in equipment or an important component, that is not available to the public and is important to the design, construction, fabrication, operation or maintenance of any sensitive nuclear facility, or any other such information that may be so designated by one of the Parties prior to its transfer under this Agreement.

ARTICLE 2

The Parties may cooperate in the field of peaceful use of nuclear energy in the following areas:

- Scientific research and development pertaining to the nuclear power sector, including nuclear reactors and their fuel cycles.

- Scientific research and development in the field of controlled thermonuclear fusion, including multilateral cooperation.

- Radioactive waste handling, decommissioning of nuclear facilities and environmental restoration.

- Nuclear and radiation safety, including issues of regulation.

- Nuclear industry and commerce.

- Shipments, based on the provisions of this Agreement, of moderator material, nuclear material, technologies and equipment, as well as services in the area of the nuclear fuel cycle, either for use in the United States of America or in the Russian Federation.

- International issues related to the peaceful use of nuclear energy, including issues of non-proliferation, IAEA safeguards, and environmental protection.

- Other areas that may be agreed upon by the Parties in writing.

ARTICLE 3

1. The Parties shall cooperate in the field of peaceful use of nuclear energy in accordance with the provisions of this Agreement and the respective legislation, regulations, norms and license requirements of the United States of America and the Russian Federation as may be applicable, and international agreements to which they are parties.

2. The Parties shall facilitate trade in moderator material, nuclear material, equipment, and technologies, as well as services pertaining to the nuclear fuel cycle, between authorized persons of the United States of America and the Russian Federation in the field of peaceful use of nuclear energy.

3. Authorizations, including import and export licenses, as well as the issuance of authorizations to third parties, relating to trade, industrial operations or nuclear material movements to the territory of the United States of America or of the Russian Federation shall not be used to restrict trade.

4. The cooperation contemplated by this Agreement as cooperation between the Parties may also be carried out between authorized persons.

ARTICLE 4

In conformity with the provisions of this Agreement, the Parties undertake to facilitate commercial relations between authorized persons of the Parties involved in cooperation in the nuclear power sector, which may include, but need not be limited to:

- investment cooperation;
- the establishment of joint ventures;
- environmental projects on an industrial or commercial scale;
- trade in nuclear material, moderator material, and relevant services.

ARTICLE 5

For the purposes of implementation of this Agreement, the Parties hereby designate the following competent authorities:

- For the United States of America, the U.S. Department of State, the U.S. Department of Energy, and the U.S. Nuclear Regulatory Commission.

- For the Russian Federation, the State Corporation for Atomic Energy "Rosatom" and the Federal Service for Environmental, Technological and Nuclear Oversight.

In case of a change in the competent authorities specified in this Article or the designation of new competent authorities, the Parties shall immediately inform each other thereof in writing through diplomatic channels, without amendment to this Agreement.

ARTICLE 6

1. This Agreement does not require the transfer of any information that the Parties are not permitted to transfer under their respective national laws and regulations, or whose transfer is inconsistent with international agreements to which the United States of America or the Russian Federation is party.

2. Restricted Data shall not be transferred by the United States of America under this Agreement.

3. Russian Federation State Secret Information as well as information similar to the information defined in paragraph 11 of Article 1 of this Agreement shall not be transferred by the Russian Federation under this Agreement.

4. The Parties recognize that they may need to protect certain information to be transferred under the terms of this Agreement by one Party to the other in connection with activities undertaken by the Government of the United States of America and the Government of the Russian Federation or on their behalf pursuant to this Agreement. In order to protect such information:

- Protected information transferred by one Party to the other shall be stamped, marked, or designated by the releasing Party as protected in accordance with its national laws and regulations. The medium in electronic, paper, or another format, containing this information, if in English, must have the marking "Protected"; if in Russian, "конфиденциально [Confidential]."

- Protected information transferred by one Party shall be protected by the recipient Party in accordance with its national laws and regulations in a manner at least equivalent to that afforded by the releasing Party. The recipient Party shall not use or permit the use of protected information for any purpose other than that for which it was transferred, and, to the extent permitted by its national laws and regulations, shall not disclose such information or transfer it to any third party not participating in the activities of the two Parties under this Agreement in connection with which the protected information was transferred, without the prior written consent of the transferring Party.

- In accordance with the laws and regulations of the United States of America, protected information transferred to the Government of the United States of America by the Government of the Russian Federation shall be treated as foreign government information transferred in confidence and shall be provided with appropriate protection from disclosure. In accordance with the legislation of the Russian Federation, protected information transferred by the Government of the United States of America to the Government of the Russian Federation shall be handled as official, restricted-distribution information and shall be provided with the appropriate protection from disclosure.

- Each Party shall limit access to protected information to persons who require access to perform a lawful and authorized government function.

ARTICLE 7

1. Nuclear material, moderator material, equipment (except for sensitive nuclear facilities, sensitive nuclear technology and major critical components) and components may be transferred for applications consistent with this Agreement.

2. Sensitive nuclear facilities, sensitive nuclear technology and major critical components may be transferred under this Agreement if provided for by an amendment to this Agreement.

3. Nuclear material may be transferred for use as fuel for reactors, in experiments, for irradiation in reactors, for enrichment to less than 20 percent in the isotope uranium-235, for conversion or fabrication, for temporary storage for purposes of further use, for use as samples, standards, detectors, targets, or for other purposes as agreed by the Parties that are consistent with the provisions of this Agreement and with the laws and regulations of the United States of America and the legislation of the Russian Federation.

4. Nuclear material, moderator material, equipment or components transferred from the territory of the United States of America to the territory of the Russian Federation, or from the territory of the Russian Federation to the territory of the United States of America, whether directly or through a third country, shall be regarded as having been transferred pursuant to this Agreement only upon confirmation, by the relevant competent authority of the recipient Party to the relevant competent authority of the supplier Party, that such nuclear material, moderator material, equipment or components will be subject to this Agreement.

ARTICLE 8

1. Plutonium, uranium-233 and high enriched uranium, transferred pursuant to the provisions of this Agreement or used in or produced through the use of nuclear material, moderator material, or equipment transferred, shall only be stored in a facility agreed upon by the competent authorities of the Parties.

2. Nuclear material, moderator material, equipment, and components transferred pursuant to this Agreement and any special fissionable material produced through the use of any nuclear material, moderator material, or equipment transferred shall be transferred only to authorized persons, and shall not be transferred beyond the territorial jurisdiction of the recipient Party unless the Parties agree otherwise.

ARTICLE 9

Nuclear material transferred pursuant to this Agreement, and nuclear material used in or produced through the use of nuclear material, moderator material, or equipment transferred, may be altered in form or content only if the Parties agree. The Parties agree that conversion, enrichment to less than twenty percent in the isotope uranium-235, fabrication of low enriched uranium fuel, irradiation or further irradiation, post-irradiation examination, and blending or downblending of uranium to produce low enriched uranium, are permissible alterations in form or content for purposes of this Agreement.

ARTICLE 10

For the purposes of implementing the rights specified in Articles 8 and 9 of this Agreement with respect to special fissionable material produced through the use of nuclear material or moderator material transferred pursuant to this Agreement, and not used in or produced through the use of equipment transferred pursuant to this Agreement, such rights shall in practice be applied to that proportion of special fissionable material produced that represents the ratio of transferred nuclear material or moderator material used in the production of the special fissionable material to the total amount of nuclear material or moderator material so used, and similarly for subsequent generations. The exact procedure for establishing the aforementioned proportion shall be agreed upon by the competent authorities of the Parties.

ARTICLE 11

1. Adequate physical protection, as specified in paragraph 2 of this Article, shall be maintained with respect to nuclear material and equipment transferred pursuant to this Agreement and special fissionable material used in or produced through the use of nuclear material, moderator material, or equipment transferred.

2. With respect to the obligation in paragraph 1 of this Article, each Party shall apply physical protection measures in accordance with its national laws and regulations at levels at least equivalent to the recommendations published in IAEA document INFCIRC/225/RevA entitled "The Physical Protection of Nuclear Material and Nuclear Facilities," and in subsequent revisions of that document accepted by both of the Parties, and the provisions of the Convention on the Physical Protection of Nuclear Material of October 26, 1979 as well as amendments to that Convention in the event of their entry into force for both Parties.

3. The Parties shall consult at the request of either Party regarding the physical protection measures maintained pursuant to this Article.

4. The Parties shall keep each other informed through diplomatic channels of those organizations or authorities responsible for ensuring levels of physical protection for nuclear material and facilities in their territory or under their jurisdiction or under their control and responsible for coordinating response and recovery operations in the event of unauthorized use or handling of nuclear material subject to this Article. Each Party shall also keep the other Party informed through diplomatic channels of the designated points of contact within its national authorized organizations for purposes of cooperation on matters involving transportation of nuclear material from the territory of its country to the territories of other countries and other matters of mutual concern.

5. The provisions of this Article shall be applied in such a manner as to avoid undue interference in the Parties' activities in the field of peaceful use of nuclear energy and to be consistent with prudent management practices required for the safe and economically justified conduct of their nuclear programs.

ARTICLE 12

Nuclear material, moderator material, equipment and components transferred pursuant to this Agreement and nuclear material used in or produced through the use of any nuclear material, moderator material, equipment or components transferred shall not be used for any nuclear explosive devices, for research on or development of any nuclear explosive devices, or for any military purpose. As specified in paragraph 9 of Article 1, military purposes shall not include provision of power for a military base drawn from any power network, production of radioisotopes to be used for medical purposes in a military hospital, and other similar purposes as may be agreed by the Parties.

ARTICLE 13

1. Nuclear material transferred to the Russian Federation pursuant to this Agreement and any other nuclear material used in or produced through the use of nuclear material, moderator material, equipment, or components transferred shall be subject, to the extent applicable, to the Agreement between the Union of Soviet Socialist Republics and the International Atomic Energy Agency for the Application of Safeguards in the Union of Soviet Socialist Republics of February 21, 1985, and the Additional Protocol that entered into force October 16, 2007 between the Russian Federation and the International Atomic Energy Agency to the Agreement between the Union of Soviet Socialist Republics and the International Atomic Energy Agency for the Application of Safeguards in the Union of Soviet Socialist Republics.

2. Nuclear material transferred to the United States of America pursuant to this Agreement and any other nuclear material used in or produced through the use of nuclear material, moderator material, equipment, or components transferred shall be subject, to the extent applicable, to the Agreement between the United States of America and the IAEA for the Application of Safeguards in the United States of America of November 18, 1977, and an Additional Protocol thereto in the event of its entry into force.

3. The Parties understand that paragraphs 1 and 2 of this Article do not require that the nuclear material referred to in those paragraphs must be in a facility that appears on the recipient Party's list of facilities that are eligible for IAEA safeguards.

4. In the event that the IAEA safeguards agreement referred to in paragraph 1 or in paragraph 2 of this Article is not being implemented, the Parties shall consult and establish a mutually acceptable alternative to that safeguards agreement consistent with their status as nuclear weapon States Parties to the NPT.

5. Each Party shall establish and maintain a system of accounting and control of nuclear material transferred pursuant to this Agreement and nuclear material used in or produced through the use of nuclear material, moderator material, equipment, or components transferred. The procedures for this system shall be those specified in the IAEA safeguards agreement referred to in paragraph 1 or 2 of this Article for the Party concerned, or, if the Parties agree, those specified in any revised version of the relevant safeguards agreement.

6. Upon the request of either Party, the other Party shall inform the requesting Party of the status of all inventories of nuclear material subject to this Agreement.

ARTICLE 14

If an agreement between either Party and another nation or group of nations provides such other nation or group of nations rights equivalent to any or all of those provided for under Article 8 or Article 9 of this Agreement with respect to nuclear material, moderator material, equipment or components subject to this Agreement, the Parties may, upon request of either of them, agree that the implementation of any such rights will be accomplished by such nation or group of nations.

ARTICLE 15

The Parties shall endeavor to avoid taking any actions that would negatively affect cooperation under this Agreement. If either Party does not comply with the provisions of this Agreement, the Parties shall promptly hold consultations on the problem, it being understood that the other Party shall have the right to temporarily suspend or to cease further cooperation under this Agreement.

ARTICLE 16

The Parties shall consult at the request of either Party regarding the implementation of this Agreement. The Parties also intend to consult regarding the development of further cooperation in the field of peaceful use of nuclear energy.

ARTICLE 17

The Parties shall consult, with regard to activities under this Agreement, to identify the world-wide environmental implications arising from such activities and shall cooperate in protecting the international environment from radioactive, chemical or thermal contamination arising from peaceful nuclear activities under this Agreement and in related matters of health and safety.

ARTICLE 18

Any dispute between the Parties concerning the interpretation or application of the provisions of this Agreement shall be promptly discussed by the Parties with a view to resolving that dispute through consultations or negotiations.

ARTICLE 19

The competent authorities of the Parties shall work out appropriate arrangements in order to effectively apply the provisions of this Agreement as they relate to nuclear material, moderator material, equipment and components subject to this Agreement. The principles of fungibility and equivalence shall apply to nuclear material subject to this Agreement. Detailed provisions for applying these principles shall be set forth in a relevant agreement.

ARTICLE 20

1. This Agreement shall enter into force on the date of the last written notification of completion by the Parties of their internal procedures necessary for its entry into force and shall remain in force for a period of 30 years. The term of this Agreement may be extended by mutual agreement of the Parties. This Agreement may be terminated by either Party by sending the relevant written notice to the other Party. In that case the Agreement shall terminate one year from the date of such notice.

2. Notwithstanding the suspension or termination, including by expiration, of this Agreement or of any cooperation hereunder, Articles 8, 9, 10, 11, 12 and 13 of this Agreement shall continue in effect so long as any nuclear material, moderator material, equipment or component subject to these Articles remains in the territory of the United States of America or the Russian Federation or under the jurisdiction or control of either Party anywhere, unless such item is no longer usable for any nuclear activity relevant from the point of view of international safeguards or has become practicably irrecoverable, or unless otherwise agreed by the Parties.

DONE at Moscow, this 6th day of May, 2008, in duplicate, each in the English and Russian languages, both texts being equally authentic.

APPENDIX E

LIST OF COMMITTEE MEETINGS AND SPEAKERS

Committee Meeting #1: June 5, 2006, Moscow, Russia
Organizational meeting

Committee Meeting #2: October 17, 2006, Washington, D.C.
Speakers
Mr. Harold D. Bengelsdorf, Bengelsdorf, McGoldrick and Associates
Dr. Phillip Finck, Argonne National Laboratory
Dr. Richard Garwin, Thomas J. Watson Research Center
Dr. Paul Lisowski, U.S. Department of Energy
Dr. Victor Reis, U.S. Department of Energy
Dr. Lawrence Scheinman, Monterey Institute of International Studies
Dr. James Timbie, U.S. Department of State
Mr. William Tobey, National Nuclear Security Administration
Dr. Frank von Hippel, Princeton University

International Workshop: April 23-24, 2007, Vienna, Austria, IAEA
Speakers
Dr. Areg Galstyan, Deputy Minister, Armenian Ministry of Energy
Dr. Karyono, Deputy Chairman for Development of Nuclear Material Cycle Technology and Engineering, National Nuclear Energy Agency (BATAN)
Mr. Tariq Rauf, IAEA
Dr. Leonan dos Santos Guimarães, Eletronuclear
Ambassador Mohamed Shaker, Vice Chairman, Egyptian Council for Foreign Affairs
Dr. Ian Smith, Executive Director and CEO, Australian Nuclear Science and Technology Organization
Prof. Dr. Jordan Stamenov, Director, Institute for Nuclear Research and Nuclear Energy, Bulgarian Academy of Sciences
Ambassador Pedro Raul Villagra Delgado, Embassy of Argentina, Canberra, Australia
Dr. Yang Chang-kook, former president, Korea Nuclear Fuel Company

Committee Meeting #3: October 9-13, 2007, Moscow, Russia
Writing meeting

Committee Meeting #4: February 12-14, 2008, Washington, D.C.
Writing meeting

APPENDIX F

JOINT COMMITTEES ON THE INTERNATIONALIZATION OF THE CIVILIAN NUCLEAR FUEL CYCLE

BIOGRAPHICAL SKETCHES

U.S. NATIONAL RESEARCH COUNCIL COMMITTEE ROSTER

John F. Ahearne, Chair, is the director of the Ethics Program at Sigma Xi, The Scientific Research Society, a lecturer in public policy at Duke University, and an adjunct scholar at Resources for the Future. His professional interests are reactor safety, energy issues, resource allocation, and public policy management. Dr. Ahearne served in the U.S. Air Force from 1959 to 1970, resigning as a major. He has also served as deputy and principal deputy assistant secretary of defense (1972-1977), in the White House Energy Office (1977), as deputy assistant secretary of energy (1977-1978), and as commissioner and chairman of the U.S. Nuclear Regulatory Commission (chairman, 1979-1981). He is a fellow of the American Physical Society, the Society for Risk Analysis, the American Association for the Advancement of Science, the American Academy of Arts and Sciences, and a member of the National Academy of Engineering, Sigma Xi, and the American Nuclear Society. From 2000 to 2003, he served as chairman of the Board on Radioactive Waste Management; he had served as a member of that board since 1993. He currently chairs the NRC Committee on Opportunities for U.S.-Russian Cooperation in Countering Radiological Terrorism, and has served on a number of other NRC committees. Dr. Ahearne holds a Ph.D.in physics from Princeton University.

Robert J. Budnitz joined the staff of the University of California's Lawrence Berkeley National Laboratory in late 2007. Before that, he was associate program leader for nuclear systems safety and security in the energy and environment directorate at the Lawrence Livermore National Laboratory. From 2002 to 2004, he directed the Department of Energy Office of Civilian Radioactive Waste Management's program on science and technology. For twenty years prior to that, Dr. Budnitz was president of Future Resources Associates, Inc. in Berkeley, California. Previously, he served as deputy director and director of the U.S. Nuclear Regulatory Commission's Office of Nuclear Regulatory Research, and he also held several management positions at the Lawrence Berkeley Laboratory of the University of California. Dr. Budnitz's professional interests are in environmental impacts, hazards, and safety analysis, particularly of the nuclear fuel cycle. He has been prominent in the field of nuclear reactor safety assessment and waste-repository performance assessment, including probabilistic risk assessment. He has served on numerous investigative and advisory panels of scientific societies, government agencies, and committees of the National Research Council. Dr. Budnitz received a B.A. degree from Yale University and a Ph.D. in physics from Harvard University.

Matthew Bunn is an associate professor in the Belfer Center for Science and International Affairs at Harvard University's John F. Kennedy School of Government. His current research interests include nuclear theft and terrorism; nuclear proliferation and measures to control it; and the future of nuclear energy and its fuel cycle. Before joining the Kennedy School in January 1997, he served for three years as an adviser to the Office of Science and Technology Policy, where he played a major role in U.S. policies related to the control and disposition of weapons-usable nuclear materials in the United States and the former Soviet Union, and directed a secret study for President Clinton on security for nuclear materials in Russia. Previously, Bunn was at the National Academy of Sciences, where he directed the two-volume study Management and Disposition of Excess Weapons Plutonium. He is the winner of the American Physical Society's Joseph A. Burton Forum Award for "outstanding contributions in helping to formulate policies to decrease the risks of theft of nuclear weapons and nuclear materials," and the Federation of American Scientists' Hans Bethe Award for "science in service to a more secure world," and is an elected Fellow of the American Association for the Advancement of Science. He is a member of the Boards of Directors of the Arms Control Association and the Partnership for Global Security. Bunn is the author or co-author of over a dozen books and book-length technical reports (most recently including Securing the Bomb 2007), and scores of articles in publications ranging from Science and Nuclear Technology to Foreign Policy and The Washington Post. Dr. Bunn holds bachelors and masters degrees in political science and a doctorate in technology, management, and policy, all from the Massachusetts Institute of Technology.

William F. Burns, Major General (USA, retired), is a former director of the U.S. Arms Control and Disarmament Agency and former commandant of the U.S. Army War College. He led the U.S. delegation on Safety, Security, and Dismantlement (SSD) of nuclear weapons, serving as ambassador in negotiations on the denuclearization of the former Soviet Union. He is a distinguished fellow at the Army War College and serves on several panels, advisory boards, and boards of trustees of governmental and non-profit organizations. He is judge emeritus of the Court of Judicial Discipline of Pennsylvania. General Burns co-chaired a National Academies study on overcoming impediments to U.S.-Russian cooperation on nuclear nonproliferation and is currently a member of the Committee on International Security and Arms Control.

Steve Fetter is dean and professor at the School of Public Policy at the University of Maryland. His research interests include arms control and nonproliferation, nuclear energy and releases of radiation, and climate change and carbon-free energy supply. He has been an advisor to many government agencies, nongovernmental organizations, and scientific organizations, and has held visiting positions at Stanford, Harvard, and MIT. From 1993 to 1994, he was a special assistant to the Assistant Secretary of Defense for International Security Policy, and in 1992 and 2004, he was a visiting fellow at the State Department. He has served on several committees for the National Academies and is currently a member of the Committee on International Security and Arms Control. He holds a Ph.D. in energy and resources from the University of California, Berkeley, and an S.B. in physics from MIT.

Rose Gottemoeller became director of the Carnegie Moscow Center in January 2006. She was previously a senior associate at the Carnegie Endowment for International Peace, specializing in arms control, nonproliferation and nuclear security issues. From 1998 to 2000, she served in the Department of Energy as Assistant Secretary for Nonproliferation and National Security and then as Deputy Under-secretary for Defense Nuclear Nonproliferation. From 1993 to 1994 she was Director for Russia, Ukraine, and Eurasia Affairs on the National Security Council in the White House. Ms. Gottemoeller co-chaired a National Academies joint consensus study on overcoming impediments to U.S.-Russian cooperation on nuclear nonproliferation and is currently a member of the Committee on International Security and Arms Control and chair of its Russia Dialogue. Ms. Gottemoeller has authored or co-authored several articles and book chapters on various aspects of nuclear nonproliferation, including *Universal Compliance: A Strategy for International Security*.

Milton Levenson is internationally recognized for his ability to apply creative new insights to major engineering challenges in the nuclear industry and for his organizational and leadership skills. Currently an independent consultant, Mr. Levenson is a chemical engineer with more than 50 years of experience in nuclear energy and related fields. His technical experience includes work related to nuclear safety, fuel cycle, water reactors, advanced reactors, and remote control. His professional experience includes research and operations positions at the Oak Ridge National Laboratory, the Argonne National Laboratory, the Electric Power Research Institute, and Bechtel. He was elected to the National Academy of Engineering in 1976. Mr. Levenson is a fellow and past president of the American Nuclear Society, a fellow of the American Institute of Chemical Engineers, and recipient of the American Institute of Chemical Engineers'' Robert E. Wilson Award in Nuclear Chemical Engineering. He is the author of more than 150 publications and presentations and holds three U.S. patents. Mr. Levenson has served as chairman or committee member for several National Academies studies and is currently a member of the Nuclear and Radiation Studies Board.

RUSSIAN ACADEMY OF SCIENCES ROSTER

Nikolay P. Laverov, co-chair, is vice president of the Russian Academy of Sciences (RAS) and former director of the Institute of Geology of Ore Deposits, Petrology, Mineralogy, and Geochemistry. He has worked in and with the USSR and Russian governments on a range of ecological problems, particularly nuclear waste disposal, and has been a leader in radiogeological studies aimed at using the protective properties of the geological environment to prevent pollution of the ecosphere by radionuclides. In addition to his research activities, Dr. Laverov has held a variety of prominent positions in scientific administration and government, including chief of the Scientific Research Organizations Administration, of the USSR Ministry of Geology (1972-1983), pro-rector of the Academy of the National Economy (1983-1987), president of the Kyrgyzstan Academy of Sciences (1987-1989), and USSR deputy prime minister and chairman of the USSR State Committee for Science and Technology (1989-1991). In 1989, Dr. Laverov was elected vice president of the USSR Academy of Sciences, a post to which he was subsequently re-elected in the RAS. In 1992, he was named co-chair of the Earth Science Joint Working Group, which is under the auspices of the U.S.-Russian Space Agreement. He is also a member of the Council on Science and Technology under the President of the Russian

Federation. Dr. Laverov graduated from the M.I. Kalinin Nonferrous Metals and Gold Institute in Moscow in 1954 and earned a doctorate in geological-mineralogical sciences in 1958. A full member (academician) of the RAS since 1987, he has authored or co-authored more than 250 publications including 20 books and has served as editor-in-chief of the journal *Geology of Ore Deposits* since 1989.

Valery S. Bezzubtsev heads the Department on Safety and Security Regulations at Nuclear Fuel Cycle Facilities at Rostekhnadzor, the nuclear regulator in the Russian Federation. Dr. Bezzubtsev coordinates regulation and inspection at nuclear energy installations including naval stations, research reactors, and nuclear fuel cycle enterprises. A 1976 graduate of the Bauman Moscow State Technical University, Department of Power Machines and Installation, he worked from 1976 through 1999 at the Scientific Research and Design Institute of Power Engineering, being involved in the development of new types of nuclear power plants. From 1999 through 2003, he served as deputy head and subsequently head of the Department for Atomic Energy at the Russian Ministry for Atomic Energy. In 2004, he was appointed as deputy head of Gosatomnadzor (GAN) before assuming his present position later that year.

Alexander V. Bychkov is director general of the Research Institute of Atomic Reactors (RIAR) in Dimitrovgrad. After graduating from Moscow State University with a degree in chemistry, he began his career at RIAR in 1982 as an engineer and researcher. He subsequently served as head of the Laboratory of Fuel Technology, head of the Fuel Cycle Department, director of the Chemical Technological Division, and deputy general director before being appointed to his current position earlier in 2006. Dr. Bychkov received his PhD from RIAR in 1998 and is a leading specialist in the field of non-aqueous methods of fuel reprocessing and a leading developer of pyroelectrochemical technologies for fast reactor oxide fuel reprocessing and production. He hold three Russian and four USSR Patents and represents Rosatom in negotiations on issues pertaining to the nuclear fuel cycle, disposal of nuclear materials, nuclear fuel reprocessing and production technologies, and investigations of nuclear fuel and new technologies.

Valentin B. Ivanov, graduated from the Samara Technical University with a degree in electrical engineering and received his doctorate of technical sciences from Kuybyshev Polytechnic Institute. His sphere of professional interests includes the nuclear fuel cycle and spent nuclear fuel management. From 1963 to 1998, he worked at RIAR, for the last nine of those years serving as its director general. From 1998 to 2002, he served as First Deputy Minister for Atomic Energy of the Russian Federation. In 2003, he was elected to the Russian State Duma, where he served as a member of the parliamentary Committee on Energy, Transport, and Communication until 2008. From 2002 to the present, he has also been employed at the RAS Institute of Ore Deposits, Petrography, Mineralogy, and Geochemistry.

Boris F. Myasoedov is deputy secretary general for science of the Russian Academy of Sciences (RAS), head of laboratories at both the RAS Vernadsky Institute of Geochemistry and Analytical Chemistry and the RAS Frumkin Institute of Physical Chemistry and Electrochemistry. His scientific activity covers such fields as the fundamental chemistry of actinides, fuel reprocessing, partitioning of radioactive waste, and environmental protection. He has authored more than 500 publications and serves as editor of the journals *Problems of Analytical Chemistry* and

Radiochemistry. Academician Myasoedov graduated from D.I. Mendeleev Chemical-Technology Institute in Moscow in 1954, and earned a PhD in radiochemistry from the Vernadsky Institute in 1965 and his full doctorate in 1975 from the same institute. He was elected to the Russian Academy of Sciences in 1994 and has been awarded two State Prizes for his research on the chemistry of transplutonium elements (1986 and 2001), the Khlopin Prize for his studies of the chemistry of protactinium (1974), and the Ipatiev Prize of the RAS Presidium in 2003.

Vladislav A. Petrov is head of the Division of Structural Petrophysics, Laboratory of Radiogeology and Radiogeoecology, of the RAS Institute of Geology of Ore Deposits, Petrography, Mineralogy, and Geochemistry. In 1984, he graduated as a mining engineer-geologist from the Moscow Geology Prospecting Institute, and he received his PhD in geology and mineralogy in 1994. For more than 20 years he has conducted field research in Russia's Krasnokamensk District, Chelyabinsk Region, and Krasnoyarsk Region on the geology of uranium deposits, radionuclide migration, site selection, and characterization of potential radioactive waste disposal sites. Dr. Petrov has made technical visits to waste disposal sites in Germany, Switzerland, France, and the United States and has authored and co-authored more than 150 articles and reports in the field of uranium geology and geological disposal of radioactive waste.

Mikhail I. Solonin is scientific director of the Technology and Innovation Center of the TVEL Corporation, which manufactures nuclear fuel. Previously, Dr. Solonin was First Deputy Minister for Atomic Energy of the Russian Federation, and director of the Bochvar State Research Center's All-Russian Scientific & Research Institute of Non-Organic Materials. His professional interests are in physical chemistry and materials science, reactor materials, technology and design of fuel rods for nuclear power plants. In 1997, he was elected associate member of the Russian Academy of Sciences (RAS corresponding member). Dr. Solonin has served as chairman of the Scientific & Technical Council (Ministry of Atomic Energy, Minatom) and as a member of Minatom Scientific & Technical Council, as a leader of the Unique Nuclear Technologies Research and Development Program, as a member of the scientific council with the Federal Science & Technology Program "Environmentally Clean Power Generation," as head of the chair on physical problems in materials science at the Moscow Physical Engineering Institute. He is a co-author of a two-volume book titled "Dispersive Fuel Rods" and a manual on nuclear reactor fuel rods for universities. Dr. Solonin graduated from the Bauman Moscow Technical University in 1968, and joined the staff of the Bochvar Institute. In 1978, he received his Ph.D. degree and in 1991, D.Sc. degree.